Electric and Plug-in Hybrid Vehicle Networks

Networks

OPTIMIZATION AND CONTROL

AUTOMATION AND CONTROL ENGINEERING
A Series of Reference Books and Textbooks

Series Editors

FRANK L. LEWIS, Ph.D.,
Fellow IEEE, Fellow IFAC
Professor
The Univeristy of Texas Research Institute
The University of Texas at Arlington

SHUZHI SAM GE, Ph.D.,
Fellow IEEE
Professor
Interactive Digital Media Institute
The National University of Singapore

STJEPAN BOGDAN
Professor
Faculty of Electrical Engineering
and Computing
University of Zagreb

RECENTLY PUBLISHED TITLES

Electric and Plug-in Hybrid Vehicle Networks: Optimization and Control,
Emanuele Crisostomi, Robert Shorten, Sonja Stüdli, and Fabian Wirth

Adaptive and Fault-Tolerant Control of Underactuated Nonlinear Systems,
Jiangshuai Huang and Yong-Duan Song

Optimal and Robust Scheduling for Networked Control Systems,
Stefano Longo, Tingli Su, Guido Herrmann, and Phil Barber

Deterministic Learning Theory for Identification, Recognition, and Control,
Cong Wang and David J. Hill

Networked Control Systems with Intermittent Feedback,
Domagoj Tolić and Sandra Hirche

Doubly Fed Induction Generators: Control for Wind Energy,
Edgar N. Sanchez and Riemann Ruiz-Cruz

Optimal Networked Control Systems with MATLAB®, *Jagannathan Sarangapani and Hao Xu*

Cooperative Control of Multi-agent Systems: A Consensus Region Approach,
Zhongkui Li and Zhisheng Duan

Nonlinear Control of Dynamic Networks,
Tengfei Liu, Zhong-Ping Jiang, and David J. Hill

Modeling and Control for Micro/Nano Devices and Systems,
Ning Xi, Mingjun Zhang, and Guangyong Li

Linear Control System Analysis and Design with MATLAB®, Sixth Edition,
Constantine H. Houpis and Stuart N. Sheldon

Real-Time Rendering: Computer Graphics with Control Engineering,
Gabriyel Wong and Jianliang Wang

Anti-Disturbance Control for Systems with Multiple Disturbances,
Lei Guo and Songyin Cao

Tensor Product Model Transformation in Polytopic Model-Based Control,
Péter Baranyi; Yeung Yam; Péter Várlaki

Fundamentals in Modeling and Control of Mobile Manipulators, *Zhijun Li and Shuzhi Sam Ge*

Optimal and Robust Scheduling for Networked Control Systems, *Stefano Longo, Tingli Su, Guido Herrmann, and Phil Barber*

Advances in Missile Guidance, Control, and Estimation, *S.N. Balakrishna, Antonios Tsourdos, and B.A. White*

End to End Adaptive Congestion Control in TCP/IP Networks, *Christos N. Houmkozlis and George A Rovithakis*

Quantitative Process Control Theory, *Weidong Zhang*

Classical Feedback Control: With MATLAB® and Simulink®, Second Edition, *Boris Lurie and Paul Enright*

Intelligent Diagnosis and Prognosis of Industrial Networked Systems, *Chee Khiang Pang, Frank L. Lewis, Tong Heng Lee, and Zhao Yang Dong*

Synchronization and Control of Multiagent Systems, *Dong Sun*

Subspace Learning of Neural Networks, *Jian Cheng, Zhang Yi and Jiliu Zhou*

Reliable Control and Filtering of Linear Systems with Adaptive Mechanisms, *Guang-Hong Yang and Dan Ye*

Reinforcement Learning and Dynamic Programming Using Function Approximators, *Lucian Busoniu, Robert Babuska, Bart De Schutter, and Damien Ernst*

Modeling and Control of Vibration in Mechanical Systems, *Chunling Du and Lihua Xie*

Analysis and Synthesis of Fuzzy Control Systems: A Model-Based Approach, *Gang Feng*

Lyapunov-Based Control of Robotic Systems, *Aman Behal, Warren Dixon, Darren M. Dawson, and Bin Xian*

System Modeling and Control with Resource-Oriented Petri Nets, *MengChu Zhou and Naiqi Wu*

Deterministic Learning Theory for Identification, Recognition, and Control, *Cong Wang and David J. Hill*

Sliding Mode Control in Electro-Mechanical Systems, Second Edition, *Vadim Utkin, Juergen Guldner, and Jingxin Shi*

Linear Control Theory: Structure, Robustness, and Optimization, *Shankar P. Bhattacharyya, Aniruddha Datta, and Lee H.Keel*

Intelligent Systems: Modeling, Optimization, and Control, *Yung C. Shin, Myo-Taeg Lim, Dobrila Skataric, Wu-Chung Su, and Vojislav Kecman*

Optimal Control: Weakly Coupled Systems and Applications, *Zoran Gajic*

Intelligent Freight Transportation, *Petros A. Ioannou*

Modeling and Control of Complex Systems, *Petros A. Ioannou and Andreas Pitsillides*

Optimal and Robust Estimation: With an Introduction to Stochastic Control Theory, Second Edition, *Frank L. Lewis, Lihua Xie and Dan Popa*

Feedback Control of Dynamic Bipedal Robot Locomotion, *Eric R. Westervelt, Jessy W. Grizzle, Christine Chevallereau, Jun Ho Choi, and Benjamin Morris*

Wireless Ad Hoc and Sensor Networks: Protocols, Performance, and Control, *Jagannathan Sarangapani*

Stochastic Hybrid Systems, *Christos G. Cassandras and John Lygeros*

Hard Disk Drive: Mechatronics and Control, *Abdullah Al Mamun, GuoXiao Guo, and Chao Bi*

Autonomous Mobile Robots: Sensing, Control, Decision Making and Applications, *Shuzhi Sam Ge and Frank L. Lewis*

Neural Network Control of Nonlinear Discrete-Time Systems,
Jagannathan Sarangapani

Fuzzy Controller Design: Theory and Applications, *Zdenko Kovacic and Stjepan Bogdan*

Quantitative Feedback Theory: Fundamentals and Applications, Second Edition, *Constantine H. Houpis, Steven J. Rasmussen, and Mario Garcia-Sanz*

Chaos in Automatic Control, *Wilfrid Perruquetti and Jean-Pierre Barbot*

Differentially Flat Systems, *Hebertt Sira-Ramírez and Sunil K. Agrawal*

Robot Manipulator Control: Theory and Practice, *Frank L. Lewis, Darren M. Dawson, and Chaouki T. Abdallah*

Robust Control System Design: Advanced State Space Techniques, *Chia-Chi Tsui*

Linear Control System Analysis and Design: Fifth Edition, Revised and Expanded, *Constantine H. Houpis, Stuart N. Sheldon, John J. D'Azzo, Constantine H. Houpis, and Stuart N. Sheldon*

Nonlinear Control Systems, *Zoran Vukic, Ljubomir Kuljaca; Donlagic, and Sejid Tesnjak*

Actuator Saturation Control, *Vikram Kapila and Karolos Grigoriadis*

Sliding Mode Control In Engineering, *Wilfrid Perruquetti and Jean-Pierre Barbot*

Modern Control Engineering, *P.N. Paraskevopoulos*

Advanced Process Identification and Control, *Enso Ikonen and Kaddour Najim*

Optimal Control of Singularly Perturbed Linear Systems and Applications, *Zoran Gajic and Myo-Taeg Lim*

Robust Control and Filtering for Time-Delay Systems, *Magdi S. Mahmoud*

Self-Learning Control of Finite Markov Chains, *A.S. Poznyak, Kaddour Najim, and*
E. Gomez-Ramirez

Nonlinear Control of Electric Machinery, *Darren M. Dawson, Jun Hun, and Timothy C. Burg*

AUTOMATION AND CONTROL ENGINEERING SERIES

Electric and Plug-in Hybrid Vehicle Networks

OPTIMIZATION AND CONTROL

Emanuele Crisostomi • Robert Shorten
Sonja Stüdli • Fabian Wirth

CRC Press
Taylor & Francis Group
Boca Raton London New York

CRC Press is an imprint of the
Taylor & Francis Group, an **informa** business

CRC Press
Taylor & Francis Group
6000 Broken Sound Parkway NW, Suite 300
Boca Raton, FL 33487-2742

First issued in paperback 2020

© 2018 by Taylor & Francis Group, LLC
CRC Press is an imprint of Taylor & Francis Group, an Informa business

No claim to original U.S. Government works

ISBN-13: 978-1-4987-4499-7 (hbk)
ISBN-13: 978-0-367-73559-3 (pbk)

Library of Congress Cataloging-in-Publication Data

Names: Crisostomi, Emanuele, 1980- author.
Title: Electric and plug-in hybrid vehicle networks : optimization and
control / Emanuele Crisostomi, Robert Shorten, Fabian Wirth, Sonja Stèudli.
Description: Boca Raton : CRC Press, Taylor & Francis Group, [2018] |
Includes bibliographical references and index.
Identifiers: LCCN 2017032134| ISBN 9781498744997 (hardback : alk. paper) |
ISBN 9781315151861 (ebook)
Subjects: LCSH: Hybrid electric vehicles--Design and construction. | Electric
vehicles--Batteries. | Structural optimization.
Classification: LCC TL221.15 .C75 2018 | DDC 629.22/93--dc23
LC record available at https://lccn.loc.gov/2017032134

Visit the Taylor & Francis Web site at
http://www.taylorandfrancis.com

and the CRC Press Web site at
http://www.crcpress.com

Contents

Preface   xiii

Acronyms   xv

1 Introduction to Electric Vehicles   1

   1.1   Introduction 1
   1.2   Benefits and Challenges 1
   1.3   Contribution of the Book 5

2 Disruption in the Automotive Industry   7

   2.1   Introduction 7
   2.2   Causes for Change 7

I   Energy Management for Electric Vehicles (EVs)   11

3 Introduction to Energy Management Issues   13

   3.1   Introduction 13
   3.2   Energy Consumption in Road Networks 13
   3.3   Distribution of Charging Facilities 14
   3.4   Interaction with the Power Grid 15

4 Traffic Modeling for EVs   17

   4.1   Introduction 17
   4.2   Traffic Model 17
      4.2.1   Basic Notions of Markov Chains and Graph Theory . 17
      4.2.2   Basic Markovian Model of Traffic Dynamics 19
      4.2.3   Benefits of Using Markov Chain to Model Mobility
          Dynamics 20
      4.2.4   Energy Consumption in a Markov Chain Traffic Model
          of EVs 21
      4.2.5   Dealing with Negative Entries 24
   4.3   Sample Applications 26
      4.3.1   Traffic Load Control 26

          4.3.1.1   Theoretical Approach   27
          4.3.1.2   Decentralized Traffic Load Control   28
  4.4   Concluding Remarks   29

5   Routing Algorithms for EVs   33

  5.1   Introduction   33
  5.2   Examples of Selfish Routing for EVs   35
  5.3   Collaborative Routing   40
      5.3.1   A Motivating Example   40
      5.3.2   Collaborative Routing under Feedback   41
  5.4   Concluding Remarks   44

6   Balancing Charging Loads   45

  6.1   Introduction   45
  6.2   Stochastic Balancing for Charging   46
  6.3   Basic Algorithm   47
      6.3.1   Charging Stations   47
      6.3.2   Electric Vehicles   47
      6.3.3   Protocol Implementation   48
  6.4   Analysis   49
      6.4.1   Quality of Service Analysis: Balancing Behavior   49
      6.4.2   Quality of Service Analysis: Waiting Times   50
  6.5   Simulations   52
  6.6   Concluding Remarks   54

7   Charging EVs   57

  7.1   Introduction   57
  7.2   EV Charging Schemes   60
      7.2.1   Control Architectures   60
      7.2.2   Communication Requirements   62
      7.2.3   Degree of Control Actuation   62
      7.2.4   Supported Services   63
      7.2.5   Control Methods   63
      7.2.6   Measurement and Forecasting Requirements   64
      7.2.7   Operational Time Scales   65
      7.2.8   Charging Policies   65
  7.3   Specific Charging Algorithms for Plug-In EVs   65
      7.3.1   Management Strategies   66
      7.3.2   Binary Automaton Algorithm   67
      7.3.3   AIMD Type Algorithm   69
  7.4   Test Scenarios   70
      7.4.1   Domestic Charging   70

	7.4.2	Workplace Scenario	70
7.5		Simulations	71
	7.5.1	Binary Algorithm	72
	7.5.2	AIMD in a Domestic Scenario	72
	7.5.3	AIMD in a Workplace Scenario	77
	7.5.4	Binary and AIMD Algorithm Scenario	77
7.6		Concluding Remarks	78

8 Vehicle to Grid — 81

8.1		Introduction	81
8.2		V2G and G2V Management of EVs	82
	8.2.1	Assumptions and Constraints	82
	8.2.2	Management of Active/Reactive Power Exchange	83
	8.2.3	V2G Power Flows	83
8.3		Unintended Consequences of V2G Operations	86
	8.3.1	Utility Functions	86
	8.3.2	Optimization Problem	88
	8.3.3	Example	89
	8.3.4	Alternative Cost Functions	90
8.4		Concluding Remarks	90

II The Sharing Economy and EVs — 91

9 Sharing Economy and Electric Vehicles — 93

| 9.1 | Introduction and Setting | 93 |
| 9.2 | Contributions | 94 |

10 On-Demand Access and Shared Vehicles — 97

10.1	Introduction	97
10.2	On Types of Range Anxiety	98
10.3	Problem Statement	99
	10.3.1 Data Analysis and Plausibility of Assumptions	100
	10.3.2 Comments on NTS Dataset	103
10.4	Mathematical Models	104
	10.4.1 Model 1: Binomial Distribution	105
	10.4.2 Model 2: A Queueing Model	106
	10.4.3 Two Opportunities for Control Theory	107
10.5	Financial Calculations	109
	10.5.1 Range Anxiety Model (VW Golf vs. Nissan Leaf)	111
	10.5.2 Range Anxiety Model with a Range of Vehicle Sizes	112
	10.5.3 Financial Assumptions and Key Conclusions	113
	10.5.4 Long-Term Simulation	114
10.6	Reduction of Fleet Emissions	116

        10.6.1 Case Study   116
    10.7 Concluding Remarks   118

11 Sharing Electric Charge Points and Parking Spaces   119

    11.1 Introduction   119
    11.2 Setting: Parking Spaces   120
    11.3 Dimensioning and Statistics   122
        11.3.1 The Dimensioning Formulae   123
        11.3.2 Parking Data and Example 124
    11.4 Efficient Allocation of Premium Spaces   129
        11.4.1 Algorithm   129
        11.4.2 Example   132
    11.5 Turning Private Charge Points into Public Ones   132
    11.6 Concluding Remarks   135

III   EVs and Smart Cities   139

12 Context-Awareness of EVs in Cities   141

    12.1 Introduction   141

13 Using PHEVs to Regulate Aggregate Emissions (twinLIN) 143

    13.1 Background   145
    13.2 Cooperative Pollution Control   147
        13.2.1 The Networked Car   148
        13.2.2 Pollution Modeling and Simulation   149
        13.2.3 Mathematical Formulation   151
        13.2.4 Integral Control   152
    13.3 Simulations   153
        13.3.1 Simulation Set-up   153
        13.3.2 Disturbance Rejection   153
        13.3.3 Extensions   155
    13.4 Concluding Remarks   156

**14 Smart Procurement of Naturally Generated Energy
    (SPONGE)   159**

    14.1 Mathematical Formulation   161
    14.2 Practical Implementation   163
        14.2.1 SPONGE Simulation Results   165
    14.3 Specific Use Case: SPONGE for Plug-in Buses   168
        14.3.1 Sponge Bus Problem Formulation   169
        14.3.2 Construction of the Utility Functions   171
            14.3.2.1 Electrical Energy Consumption   171

14.3.2.2 Saving of CO_2 171
14.3.2.3 Utility Functions f_i 172
14.4 Optimization Problem 173
14.5 Simulation Results 175
14.6 Concluding Remarks 176

15 An Energy-Efficient Speed Advisory System for EVs   **179**

15.1 Introduction 179
15.2 Power Consumption in EVs 180
15.3 Algorithm 182
15.4 Simulation 185
15.4.1 Consensus and Optimality 185
15.5 Concluding Remarks 186

IV   Platform Analytics and Tools   **191**

16 E-Mobility Tools and Analytics   **193**

16.1 Introduction 193

17 A Large-Scale SUMO-Based Emulation Platform   **195**

17.1 Introduction 195
17.2 Prior work 196
17.3 Description of the Platform 198
17.4 Sample Application 201
17.5 Concluding Remarks 201

18 Scale-Free Distributed Optimization Tools for Smart City Applications   **205**

18.1 Introduction 205
18.2 The AIMD Algorithm 205
18.3 Optimal Resource Allocation 207
18.4 Scale-Free Advantages of AIMD 209
18.5 Passivity 210
18.6 Concluding Remarks 212

Postface   **213**

References   **215**

Index   **235**

Preface

This book describes work carried out by the authors and their co-authors during the period 2011-2017. From its beginnings at the Hamilton Institute in Ireland, the work eventually embraced a rich network of researchers in several disciplines from across the globe, involving collaborators from North America, Europe, Australia and Asia, and researchers from both academia and industry.

The period 2011-2017 will probably be considered to be a very disruptive period in the evolution of the automobile. Cars have basically been in the same form, with the same functionality, since the invention of the diesel engine. Now, suddenly, disruption and innovation are coming from every direction, causing a rethink of the ways that cars are designed and used in cities. It is our great fortune to have been active in the automotive area during this period, and to have been able to work on some of the research challenges that have arisen.

As we have mentioned, this book describes work carried out not only by the authors, but also by a host of other collaborators, to all of whom we owe a huge debt of gratitude.

First and foremost, we would like to thank our Ph.D. and Masters students who worked directly on this topic. In particular, we would like to mention and acknowledge the contributions of Arieh Schlote, and Florian Hausler who were (along with Sonja) our first students working on this topic, as well as the more recent contributions of Mingming Liu, Yingqi Gu, and Eoin Thompson. All of the aforementioned contributed greatly to our EV work, and many of our joint results are reported in this book.

Thanks is also due to our colleagues, Wynita Griggs and Rodrigo Ordóñez-Hurtado, for their substantial contributions.

We are also greatly indebted to our close collaborators: Chris King (Northeastern University); Martin Corless (Purdue University); Jia Yuan Yu (Concordia University); Joe Naoum-Sawaya (Ivey Business School); Giovanni Russo, Jakub Mareček (both IBM Research); Kay Massow, Ilja Radusch, Thomas Hecker (all from Fraunhofer Fokus); Steve Kirkland (University of Manitoba); Rick Middleton (the University of Newcastle, Australia); Astrid Bergmann and Jörg Raisch (both Technical University of Berlin);

Julio Braslavsky (CSIRO Energy); Mahsa Faizrahnemoon (Simon Fraser University); and Brian Purcell (Nissan Ireland). Finally we thank Julian Danner for his tireless work in helping prepare the figures in this book.

Robert Shorten also thanks ESB swimming club for facilitating work on this manuscript during the long winter training sessions of 2016-17.

We are also very grateful to our funding agencies; in particular Science Foundation Ireland.

Finally, we thank CRC Press -Taylor & Francis for giving us the opportunity to write this book. In particular, we would like to thank Nora Konopka for supporting this project, Kyra Lindholm for coordinating the manuscript preparation, Karen Simon for handling the final production, Shashi Kumar for his LaTeX assistance, and John Gandour for designing the book cover.

References and footnotes
Throughout this book we shall use references for archival sources, and footnotes, among other things, for non-archival material such as websites. Some figures and partial content are reprinted with permission from published papers. Where appropriate, chapters containing such content are indicated using footnotes: ©*IEEE. Reprinted, with permission, from* source.

MATLAB® is a registered trademark of The MathWorks, Inc. For product information please contact:
The MathWorks, Inc.
3 Apple Hill Drive
Natick, MA, 01760-2098 USA
Tel: 508-647-7000
Fax: 508-647-7001
E-mail: info@mathworks.com
Web: www.mathworks.com

Acronyms

ADAS   Advanced Driver Assistance System.

AI   Additive Increase.

AIDL   Android Interface Definition Language.

AIMD   Additive Increase Multiplicative Decrease.

API   Application Programming Interface.

CCCV   Constant Current, Constant Voltage.

DER   Distributed Energy Resource.

EMU   Engine Management Unit.

EV   Electric Vehicle.

EVRP   Electric Vehicle Routing Problem.

FEV   Full Electric Vehicle.

G2V   Grid to Vehicle.

GMFV   Guaranteed Minimum Future Value.

HEV   Hybrid Electric Vehicle.

HIL   Hardware-in-the-Loop.

HVAC   Heating, Ventilation, Air Conditioning.

I2V   Infrastructure to Vehicle.

ICE   Internal Combustion Engine.

ICEV   Internal Combustion Engine Vehicle.

IoT   Internet of Things.

ISA   Intelligent Speed Advisory.

ITS   Intelligent Transportation Systems.

JOSM   Java OpenStreetMap Editor.

KKT   Karush-Kuhn-Tucker.

MD   Multiplicative Decrease.

MFPT   Mean First Passage Time.

NMAE   Normalized Mean Absolute Error.

NTS   National Travel Survey.

OEM   Original Equipment Manufacturer.

PHEB   Plug-in Hybrid Electric Bus.

PHEV   Plug-in Hybrid Electric Vehicle.

PV   Photovoltaics.

QoS   Quality of Service.

SOC   State Of Charge.

SPONGE   Smart Procurement Of Naturally Generated Energy.

SUMO   Simulation of Urban MObility.

TCP   Transmission Control Protocol.

TraCI   Traffic Control Interface.

V2G   Vehicle to Grid.

V2I   Vehicle to Infrastructure.

V2V   Vehicle to Vehicle.

V2X   Vehicle to Infrastructure and Vehicle to Vehicle.

VANET   Vehicular Ad-hoc NETwork.

YoY   Year on Year.

1

Introduction to Electric Vehicles

1.1 Introduction

Growing concerns over the limited supply of fossil-based fuels are motivating intense activity in the search for alternative road transportation propulsion systems. In addition, regulatory pressures to reduce urban pollution, CO_2 emissions and city noise have made plug-in electric vehicles [23, 166] a very attractive choice as the alternative to the internal combustion engine [140]. However, despite the enormous benefits of such vehicles, their adoption and uptake has, to this point, been disappointing. In this chapter we shall outline some of the impediments to electric vehicles, and discuss some of the solutions to these problems that will be addressed in this book, as well as other opportunities that arise when using this new form of mobility.

1.2 Benefits and Challenges

Basically, an Electric Vehicle (EV) is a vehicle that no longer relies solely on an Internal Combustion Engine (ICE) as the only propulsion mechanism, but rather uses an electric drive system as a replacement, or to enhance, the ICE. Roughly speaking, three types of electrically propelled vehicles can be distinguished.

- A Hybrid Electric Vehicle (HEV) combines an ICE and an electric motor within the drive train. Mostly, the electric motor supports the ICE for fuel economy and/or performance. The vehicle is then either propelled by the combustion engine or the electric drive.

- A Plug-in Hybrid Electric Vehicle (Plug-in Hybrid Electric Vehicle (PHEV)) is a vehicle equipped, in general, with a larger battery compared to HEVs, that allows recharging of the battery via home outlets or at charging stations. While in most cases both the electric drive and the ICE are able to propel the vehicle, some vehicles use solely the electric drive. In this latter case the ICE can be used to recharge the battery or directly produce electricity for the electric drive. Also, in most cases PHEVs can be used in a full electric mode if there is enough energy stored in the battery. This allows one to select when and where to release pollutants. This functionality shall be used in some applications discussed in the book.

- A Full Electric Vehicle (FEV) runs solely on an electric drive system. As with PHEVs their batteries are large and can be recharged in charging stations or at home. Since there are no pollutants released while driving, these vehicles are often marketed as zero-emission vehicles. Naturally, this is not exactly a correct terminology, since the recharging of the batteries will cause emissions depending on the actual emissions of the power generation in the country. Due to the fact that many power plants are located in less populated areas, the use of FEVs still has beneficial effects on emissions in population centers. Such vehicles may be considered *as filters for turning dirty into clean energy.*

Of these three types, we shall distinguish PHEVs and FEVs from HEVs, and we shall denote the former as plug-in EVs, to emphasize that they continuously have to recharge their batteries. In Figure 1.1 a graphical overview over the various EV types is given.

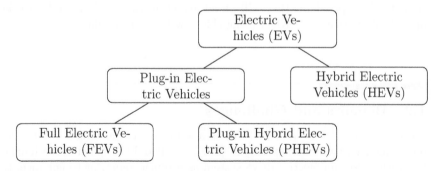

FIGURE 1.1
Classification of some different EV types

While the deployment of plug-in EVs can give rise to various environmental and health improvements, their adoption to date has been disappointing. According to initial reports [6], even in Europe, where the green agenda was well received, fewer than 12 000 EVs were sold in the first half of 2012 (of which only 1000 of these were sold in the UK). This number represented less than 0.15% of total new car sales in that year. These figures were in spite of the fact that many European governments had offered incentives for the purchase of EVs in the form of subsidies and had also invested in enabling infrastructure. There are however hints that the numbers may soon dramatically change. For instance, growing 59% year over year (YoY), approximately 12 000 electric cars were sold across the US in January 2017, accounting for approximately 1% of US auto sales[1]. Similarly, the Chinese market had more than 32 000 new electric cars on the streets in March 2017, an 89% increase over the same month the previous year, with the annual growth rate at 31%[2]. Numbers in Europe remain contradictory, with Northern countries leading the market (e.g., in 2017 Norway has the highest per capita number of all-electric cars in the world: more than 100 000 in a country of 5.2 million people[3]). However, things seem to be speeding up, recently Volvo have announced that all new cars launched from 2019 onwards will be partially or completely battery-powered. The company called this step a "historic end" to building models that only have an internal combustion engine[4]. The day after the Volvo announcement, Emmanuel Macron's government announced that France will end sales of petrol and diesel vehicles by 2040, as part of an ambitious plan to meet the targets of the Paris climate accord[5]. Only few days later, a similar plan was unveiled to ban the sale of new diesel and petrol cars by 2040 in the UK in a bid to encourage people to buy electric vehicles[6].

Despite such recent promising signals, still the percentage of traveling EV remains very low at a global scale. Some of the main factors hindering the widespread adoption of EVs from the point of view of customers are as follows:

1. **Price:** EVs have, to date, been expensive, even when subsidized. A major factor in the cost of such vehicles is the cost of the battery [13]. While battery costs are forecast to reduce dramatically over the next few years [27, 158], this is currently an important aspect in understanding the sales of EVs. In response to this, some companies, are proposing to lease batteries to the customer to offset some of the battery related costs.

2. **Vehicle size:** EVs are sometimes small with limited luggage space to reduce energy consumption, or to accommodate batteries (in some hybrid vehicles). This is sometimes a problem for potential purchasers of vehicles who, on occasion, would like to transport significant loads using their vehicles.

3. **Long charging times:** Charging times for plug-in EVs can be long [201]. An often cited fact by advocates of electric vehicles in response to this is that fast charging methods can service average vehicles in about 30 minutes [35, 28]. Such time-scales may be just about acceptable to a normal

---

[1]https://cleantechnica.com/2017/02/04/us-electric-car-sales-59-january-2017/. Last Accessed July 2017.

[2]https://evobsession.com/china-electric-car-sales-keep-soaring-march-2017/. Last Accessed July 2017.

[3]http://e360.yale.edu/features/with-norway-in-the-lead-europe-set-for-breakout-on-electric-vehicles. Last Accessed July 2017.

[4]https://www.theguardian.com/business/2017/jul/05/volvo-cars-electric-hybrid-2019. Last Accessed July 2017.

[5]https://www.theguardian.com/business/2017/jul/06/france-ban-petrol-diesel-cars-2040-emmanuel-macron-volvo. Last Accessed July 2017.

[6]http://www.telegraph.co.uk/news/2017/07/25/new-diesel-petrol-cars-banned-uk-roads-2040-government-unveils/. Last Accessed July 2017.

car owner. However, in the presence of queuing, 30 minutes can rapidly become several hours, and push such fast charging stations into the realm of "not acceptable". Thus, it is likely that overnight or workplace charging will be the principal method of vehicle charging for the foreseeable future. An associated issue in large cities concerns the availability of charging points. This is especially an issue in cities with large apartment block type dwellings.

4. **Limited range:** Maximum ranges of less than 300 km in favorable conditions are not unusual for EVs, and this reduces significantly when air-conditioning or heating is switched on [49]. Hence, the range is not only limited but to a certain degree also unpredictable, which worsens the issue. Additionally, other issues, that are a nuisance for normal ICE vehicles, are exacerbated as a result of the limited range. For example, the cost of searching for a parking space at the end of a journey is much higher than for a conventional vehicle, because the EV's range is low and therefore energy should not be wasted searching for a parking spot. Research is ongoing to address these issues, with much of the current work focusing on new battery types, optimal vehicle charging, vehicle routing, and in-vehicle energy management systems with a view to minimizing wastage of energy and thereby increasing vehicle range [162]. [7]

The latter two issues are often grouped together as one and discussed under the title of *range anxiety* [140, 180]. Further challenges that arise include the following.

1. **Charging (from the perspective of generation distributors)**: The energy that will potentially be required to charge the large volumes of batteries of EVs will considerably increase the load on the distribution grid, and can cause power quality issues when not regulated.

2. **Traffic management:** While traffic management in general is an important factor, the issue becomes more pressing if EVs are present due to their limited energy availability. Hence, a traffic jam or rerouting due to road work or accidents, can have a strong impact on the energy required for the journey and in the worst case force the drivers to recharge before the end of the journey.

3. **Charging Infrastructure:** While momentarily most charging occurs at home during night-time, an important consideration is the availability and distribution of charging stations. This is related to the issue of limited

---

[7]Some of these issues lead to changes in driver behavior when faced with the need to increase range. For example, in [194], behavioral adaptations (in response to limited available energy) were observed among participants of a study group, who were leased a battery EV for a year. Some of these behavioral adaptations included turning off the air conditioning or heater and driving more slowly, as well as swapping vehicles with other users.

range and long charging times, i.e. range anxiety. A major issue in this context is charge point anxiety - that is the angst associated with not being able to access a charge point when needed.

4. **Electromagnetic emissions:** Another issue regarding EVs concerns electromagnetic emissions. While there is no evidence that electromagnetic radiation from EVs is dangerous, this issue is a focus point for regulatory authorities (see e.g. the EU Green Car Programme) and has been raised by several research agencies [7].

5. **Battery related issues:** A further concern is whether enough lithium can be sourced to build batteries to construct enough vehicles to replace the existing passenger vehicle fleet. Are we simply substituting one rare resource (oil) with another (lithium)? Also, the transportation of batteries is not trivial and necessitates special precautions [126, 9]. Finally, most reasonably sized batteries are not capable of realizing the range enjoyed by conventional ICE based vehicles, which comes in play in regard to the above mentioned issue of range anxiety. While this latter issue is the subject of much research, battery size and performance currently represent one of the major determinants in the design of EVs today [14, 13].

1.3   Contribution of the Book

Our objective in this book is to address some of the issues that impede the adoption of plug-in EVs. Rather than focusing on single vehicles, our focus shall be on developing techniques to better use networks of electric vehicles. We believe, at the time of writing this book, that this aspect of EV technology has not been significantly documented elsewhere.

To this end we partitioned the book into four parts. The first part of the book is concerned with energy management. Topics that we shall consider include: plug-and-play infrastructure for charging fleets of vehicles; how energy is dissipated in electric vehicles; how to avoid queuing at charging stations; routing of EVs to consume energy efficiently; and a consideration of some of the unintended consequence of plug-in EV usage.

The second part of the book will consider using ideas from the sharing economy to better share "road electrification". Topics that we shall consider include: on demand mobility for EVs; and the sharing of personal charge points.

The third part of the book will focus on the actuation possibilities afforded by the use of PHEVs. By judiciously selecting when/where one engages the electric motor, a range of new ideas can be implemented in cities. Specifically, we shall see how such ideas can be used to regulate emissions in a local region, better balance the needs of the grid and the transportation network, and manage energy consumption in a fleet of vehicles.

Finally, the last part of the book discusses analytics that can be used to support the design and testing of electro mobility (E-mobility) concepts without the need for large scale fleet testing.

2

Disruption in the Automotive Industry

2.1 Introduction

This is one of the most exciting times for working in the automotive industry. In the past few years, many disruptive forces have emerged and these are empowering real change in the way cars are sold, used and conceived. Driven by examples of companies in other industry branches that did not respond to disruptive technologies, most notably Kodak [55], automotive Original Equipment Manufacturers (OEMs) are currently embracing these new technologies and searching for new ways to deliver mobility to consumers, and to monetize mobility platforms. It is in the context of this changing landscape that this book is written.

2.2 Causes for Change

At this present time change in the automotive industry is being driven by a number of forces. At a very coarse level, these can be categorized as follows (in no particular order).

(i) **Connectivity:** We are well on the way to a point where vehicles can communicate seamlessly with each other and with the road infrastructure in place. This is creating new opportunities for services for drivers, passengers, city managers and general citizens and is opening up new vistas of service delivery that can be monetized by OEMs.

(ii) **New vehicle types:** Important changes have also taken place in the types of vehicles that are used. Gone are the days when the *Internal Combustion Engine (ICE)* ruled supreme. New vehicle types, such as EVs, HEVs, PHEVs, Fuel Cell vehicles, and electric bikes, are becoming more common. These vehicles are not simply replacements for the ICE, they should be seen as new forms of mobility, and offer new methods of actuation for regulators to combat pollution and manage energy consumption [164, 88, 80]. The HEV is one such vehicle type. As we shall see in this book, by orchestrating the switching between electric and ICE mode, several problems in cities can be addressed. Similarly, by viewing the battery in EVs as a filter for turning dirty energy into clean energy, huge opportunities arise in the manner in which EVs can be deployed.

(iii) **Algorithmic developments:** An important aspect of *smart mobility*, often forgotten, is that much of the heavy lifting required for real practical progress, has been completed by the *networking* community. Many of the mathematical boundaries in networking, associated with large scale distributed control and optimization, have been pushed back by this community, to a point where very large scale distributed solutions can be implemented over graphs with time-varying connectivity properties. Perhaps the best example of such a contribution is the TCP/IP protocol. As we shall see, the *Additive Increase Multiplicative Decrease (AIMD)* algorithm, developed as part of this protocol, can also be used to orchestrate and coordinate fleets of vehicles in an optimal manner. Importantly, this can be done without the need for inter-vehicle communication, and with only intermittent feedback from a central coordinator.

(iv) **Demographic changes:** An important driving force in the automotive industry is being caused by changing demographics. Put simply, there is a discernible trend emerging among younger generations away from traditional car ownership models, towards an *on-demand* model. This is creating new opportunities for OEMs in the way cars are used (car and ride sharing), and creating the need for new financial models for car payment.

(v) **Platform monetization:** Companies such as *Apple* have pioneered a vision that has moved computers from being simple computing devices, to being more general delivery platforms. A similar, and potentially more profitable journey is now underway in the automotive industry. Cars, because of their physical size, and because consumers, when in these vehicles, are captive, have tremendous potential for delivering auxiliary services to drivers and passengers alike. An indication of the size of this market can be seen from the amount of radio advertising that is delivered in-car. This is a huge missed opportunity for OEMs. They provide the delivery platform, yet derive none of the income from this revenue stream. There is now a concerted effort from OEMs to avoid this situation arising again in the future.

(vi) **By-products of the ICE and aggregation:** Another force that is driving change in the automotive sector is the realization that the ICE is causing problems for humans in three distinct ways. First, there is an irrefutable connection between road transportation and global warming. Second, there is now a growing realization that the ICE and its by-products (particulate matter, ozone, benzene, nitrate oxides, carbon monoxide), are harmful to human health and are having negative consequences for air quality in many cities. Few would knowingly swim in a dirty swimming pool. Yet, on a daily basis, we do something very similar with the air that we breathe. There has been considerable regulatory pressure in many regions of the world to improve air quality. The response from the automotive industry has been to make cars cleaner, if only at the test stand and less so in real operation. However, even with effective reductions of emissions from individual cars, as car volumes continue to grow, *the aggregate effect* may give rise to an increase in pollution rather than a decrease. Thus, the third factor associated with the ICE, is the realization that even without the connected car, people are connected through the air they breathe, and it is the aggregate behavior that affects the quality of this resource.

(vii) **Regulation:** Another force for change in transport is regulation. Regulation is driven by four main considerations: safety; greenhouse gases; congestion; air quality. Much of the currently proposed regulation is concerned with air quality and proposes severely limiting ICE access to cities from the near future[1] [153].

(viii) **Partial and full autonomy:** The penultimate driving force in transportation is the seemingly unstoppable march towards autonomous driving. Full, and partial autonomy, are real topics of interest for OEMs. The marriage of vehicle electrification, and full and partial autonomy, is likely to give rise to significant innovation opportunities in the near future.

(ix) **Sustainability and constrained resources:** The final driving force is related to the aggregation effect mentioned in (vi). Throughout society, there is a move away from the assumption of infinite resources, to an assumption of a contained resource. This gives rise to economic concepts of budgets, sharing, and utility maximization. The development of protocols for sharing a restricted resources among competing actors in way that maximizes overall utility for the group of users is one of the challenging tasks to be solved in the context of future mobility.

Much of the work that we shall describe is motivated by some of the forces described above. Our particular perspective is on the networks of EVs and how these can be orchestrated in a manner that derives maximal benefit for the user of the vehicle and for society. As we shall see, this perspective shall lead to a rich exploration involving distributed control and optimization, queueing theory, as well as some new ideas in electro-mobility.

---

[1]http://www.independent.co.uk/environment/climate-change/norway-to-ban-the-sale-of-all-fossil-fuel-based-cars-by-2025-and-replace-with-electric-vehicles-a7065616.html. Last Accessed July 2017.

Part I

Energy Management for EVs

3

Introduction to Energy Management Issues

3.1 Introduction

In this part we will investigate issues that are linked to the energy management of EVs. This area includes the management or distribution of energy for a single vehicle, as well as for networks of vehicles. Much of the current work in this area is focusing on new battery types, optimal charging, and in-vehicle energy management systems. On the other hand, energy management on an aggregated scale, though rarely treated in the literature, is critical if many EVs are used. In this part we will mainly focus on the latter case.

Energy management issues on an aggregated level can broadly be categorized in three ways: energy consumption with respect to travel on a road network; the usage and distribution of charging facilities; and the interaction of vehicles with the power grid. We will here briefly introduce the main issues and concepts in these three areas.

3.2 Energy Consumption in Road Networks

Intelligent traffic management is essential to achieving two main goals: the reduction of harmful emissions; and the improvement of the efficiency of the transportation network. A key enabling technology in developing traffic management strategies is the availability of accurate traffic models that can be easily used for both prediction and control. A major objective in developing such models is the development of smart traffic management systems that proactively predict traffic flows and take pre-emptive measures to avoid incidents (traffic build up, pollution peaks etc.) rather than reacting to traffic situations [40]. Such measures include: 1. recommending alternative vehicle paths based on advanced routing techniques; 2. adjusting the timing and the phasing of green periods in traffic lights; and 3. changing the speed limits, or the recommended speeds.

Recent examples in the literature that implement such actions can be found in [79], [142] and [122] respectively. In the case that a large portion of the traffic consists of plug-in EVs, especially of FEVs, the benefits of such smart traffic management systems are even greater than for conventional vehicles. The main reason for this is the dependence of journey fidelity on consumed energy over each route, and the relatively long charging times that are needed once the vehicle battery is depleted. For instance, congestion in one area of the city can give rise to longer queues at a charging station within that area when the plug-in EV owners try to charge. Similarly, routing and rerouting of EVs will affect the consumed energy of a given vehicle. Hence, it is important to give the users the option to choose energy consumption as an optimality criterion, in a similar way as the user now can select, for example, between the shortest path (e.g., in km), the minimum time path, the most economic path (e.g., where toll charges are minimized) and the minimum fuel path (i.e., where fuel consumption is taken into explicit account). Additionally, the constraints due to the limited battery capacity and the charging options should ideally also be taken into account during route planning, which consequently requires a complete redesign of the common routing strategies. Thus, effective traffic management is of utmost importance to tackle many of the issues hindering the adoption of EVs. To achieve this goal, the availability of accurate traffic models is recognized as a compelling topic in the context of EVs. However, very few examples of such dedicated bespoke traffic models exist in the literature for EVs. Notable exceptions are [96] and [162]. The latter model is of use in the context of this book and will be presented in detail in Chapter 4. Note that accurate traffic models also assist the electricity grid by supporting the power providers with accurate predictions of the energy demand.

3.3   Distribution of Charging Facilities

One of the main factors associated with range anxiety is linked to the long charging times of FEVs [24]. While long time scales are acceptable for charging at home or at work, they become prohibitive once the range is insufficient for the journeys made between charging periods. For charging during the journeys, even fast charging, which takes 15 minutes or more, is unacceptable once the need for queuing at charging stations is taken into account, which can prolong charging times up to an hour or longer. Thus, the benefits of optimizing routing and in-vehicle systems to maximize the range of FEVs, appears of minor impact if all of the saved energy is used while searching for convenient charging opportunities.

The importance of developing efficient methods to prevent FEVs from wasting time while queuing at charging stations is also illustrated by the volume of recent published works on the topic. Among these, we note [50], where vehicles compute a routing policy that minimizes the expected journey time while considering the *intentions* of other vehicles (i.e., intentions of charging en route); and [30], where a reservation-based charging approach is proposed.

In particular, in these and other works, reservations are periodically updated for requesting the change of the charging station selection decision to take into account the uncertain aspects of traffic; [82], where a higher level distributed scheduling algorithm together with a lower level cooperative control policy for individual FEVs on a highway is designed to optimize the operation of the overall charging network; [58], where charging operators of fast charging stations can set a limit on the FEV's requested State Of Charge (SOC) in overloaded conditions to increase total charged energy, revenues for the charging stations as well as decrease the probability of blocking arriving FEVs. An alternative approach to this problem is described in Chapter 6 where the load balancing approach described in [87] is given (this can be viewed as a kind of dual of the routing problem). Note also that the collaborative routing strategies in Chapter 5 tackle the load balancing problem by actively taking charging into account.

3.4   Interaction with the Power Grid

Even though the adoption of plug-in EVs is slower than expected, the future uptake of these vehicles will add a non-negligible energy load on the power grid, due to the charging process. In this context a high penetration of plug-in EVs may increase the stresses on the grid significantly and cause, in extreme cases, problems such as voltage deviations, line overloading, or transformer overloading [155, 146, 151, 85, 37]. Based on anticipated uptake levels ranging from 10% up to 50%, studies agree that these adverse effects limit the amount of plug-in EVs that can be charged simultaneously without additional investments for the grid.

Another method to handle the extra charging load by the plug-in EVs is to control the charging procedure such that the adverse effects on the electricity grid are minimized (i.e., the EVs participate in a load management scheme). There are two main reasons for this being a viable alternative. Firstly, the electricity grid has to be constituted to deal with the peak power demand; hence there are long periods during which the electricity grid has spare capacity available. Using this spare capacity for charging will reduce the stresses on the grid, while at the same time increase its efficiency. Secondly, plug-in EVs are ideal candidates for load management due to their highly flexible demand elasticities. We will in Chapter 7 introduce the concepts of load management and give a brief review over proposed methods to achieve controlled charging. Also, in Chapter 7 we will discuss in detail a load management scheme that was previously proposed in [172, 171, 176]. This management scheme deals with the simultaneous charging of vehicles over longer periods, mostly during night time or work place charging.

In addition, many works also discuss the use of plug-in EVs to support the distribution grid with energy in times of need, since their internal battery allows them to act as a storage device. This is often termed Vehicle to Grid (V2G) operation and leads to various benefits for the distribution grid. In Chapter 8, we explain how the scheme proposed in Chapter 7 can be extended to allow for such operations. In this context, we will also discuss the consequences that V2G operations can have on the environment and illustrate why these should be considered.

Notes and References

Energy management is a central issue for EVs both in terms of the individual energy management and the aggregated effects in traffic and during charging. The control mechanisms in place not only influence a single vehicle owner but cause interactions with other road users as well as any electricity consumer. The remainder of this part deals with issues associated with these topics.

This part of the book is based on the following papers by the authors and their co-workers. The chapter on traffic modeling for EVs is based on a paper written in collaboration with Arieh Schlote and Steve Kirkland [162]. The chapter on routing is based on several works of the authors (such as [162] in collaboration with Steve Kirkland and Arieh Schlote, [42] in collaboration with Steve Kirkland, [163] in collaboration with Arieh Schlote[1]). The chapter on load balancing is based on work [87] in collaboration with Florian Häusler, Arieh Schlote and Ilja Radusch[2]. The chapters on charging mechanisms are mainly based on works [172] and [169] in collaboration with Rick Middleton. Finally, the chapter on V2G energy exchange is based on works [173] and [175] in collaboration with Rick Middleton and Wynita Griggs respectively[3].

---

[1]©IEEE. Figures 5.3 and 5.4 reprinted, with permission, from [163].
[2]©IEEE. Reprinted, with permission, from [87].
[3]©IEEE. Reprinted, with permission, from [175].

4

Traffic Modeling for EVs

4.1 Introduction

This chapter reviews currently existing traffic models for EVs with particular emphasis on a recently introduced Markov chain based model. In addition, some applications based on this traffic model are described. Very few examples in the literature of traffic models exist for EVs. Notable among these are [96] and [162]. While both of them are based on the use of Markov chains, [96] captures the diurnal variation in the use of a single vehicle. On the other hand, [162] provides a macroscopic description of mobility flows. In this chapter we are interested in this second model, as it has been proved useful in many application studies to accurately describe an aggregate behavior of a network of vehicles. This model is based on homogeneous Markov chains, and adapts the model initially proposed for conventional vehicles (see [40]), which in turn was similar in spirit to the well-known Google's PageRank algorithm [114]. The remainder of this chapter is organized as follows: Section 4.2 describes the traffic model for EVs introduced in [162]; Section 4.3 describes a number of traffic applications that can be planned on the basis of the proposed model; finally, Section 4.4 concludes the chapter and outlines interesting lines of research to further extend the work presented here.[1]

4.2 Traffic Model

4.2.1 Basic Notions of Markov Chains and Graph Theory

First we recall some basic notions on Markov chains, as they underpin the EV traffic model that we are going to discuss here. A Markov chain (MC) is a

---

[1]Note that mathematical models are not the only way to represent traffic and tackle the aforementioned problems. Other authors have preferred to simulate traffic in a city, and evaluate how traffic would react to single control actions. See for instance [130] and [71] for an example of this, where traffic simulated in Simulation of Urban MObility (SUMO) was used with the ultimate goal of computing efficient routing strategies and for dimensioning the number of public charging stations in an urban context respectively.

discrete time stochastic process with a finite or countable number of states. At each time step, the state changes and the new state is chosen in a probabilistic manner. The transition probabilities depend only on the state of the chain at the previous time step and not on the past history of the process. Throughout this chapter we shall only consider homogeneous Markov chains. We denote by p_{ij} the probability of going from state i to state j in one time step and if the number of states is finite, then the matrix $\mathbf{P}$ of elements p_{ij}, together with the initial distribution vector, fully describes the evolution of the Markov chain. The matrix $\mathbf{P}$ is called the transition matrix and it is a row-stochastic non-negative matrix, as the elements of each row are probabilities and they sum up to 1.

There is a strong link between Markov chains with finite state space and graphs. A graph is a set of states (or nodes or vertices) that can be connected by edges (or arcs). We will only consider directed edges here. An edge from state i to state j indicates that it is possible to make a transition from state i to state j. It is possible to give a weight to each edge that corresponds to the cost of using that edge. If the aggregate cost of all outgoing edges of each node is normed to 1 then the costs can be interpreted as the probabilities of using the corresponding edges. In this context states of a given Markov chain can be associated with nodes in the graph and non-zero probabilities of transition between two states in the chain can be associated with directed edges between the corresponding nodes with the given probability as a weight.

The graph is called strongly connected if starting from any node it is possible to reach any other node by following the edges. The graph is strongly connected if and only if the transition matrix of the corresponding Markov chain is irreducible. Throughout this chapter and the remainder of this manuscript, we will assume that all the Markov chain transition matrices considered are irreducible and primitive unless noted otherwise. We can apply the Perron-Frobenius theorem, see for example [94, Theorem 8.8.4], to ensure the existence of an invariant measure π with $\pi^{\top}\mathbf{P} = \pi^{\top}$. In this situation, π is entry-wise positive and its entries sum to 1; we call π the chain's stationary distribution.

An important notion in the study of irreducible Markov chains is the Mean First Passage Time (MFPT) m_{ij}, which gives the expected number of steps of a random walk starting from vertex i and finishing in vertex j governed by the weights of the graph's edges [102]. A related concept to the MFPTs is given by the Kemeny constant, which is the expected cost of a random trip on the graph where the destination is chosen according to the stationary distribution. For node i this value can be computed as

$$K_i = \sum_{j \neq i} m_{ij}\pi_j. \tag{4.1}$$

It is well known that K_i is independent of the starting node i, and thus is a global parameter for the Markov chain [102], that in some cases may be used as a global efficiency measure of the road network (e.g., see [40], [57]).

4.2.2 Basic Markovian Model of Traffic Dynamics

The utilization of Markovian models to represent traffic dynamics goes back to the paper [40]. We now recall the main concepts behind this model, and then we explain how it can be extended to describe EV dynamics.

Graphs and Markov chains can be used quite naturally to model urban traffic networks. The starting point of the Markovian model describes the transitions between road segments. The directed graph associated with the

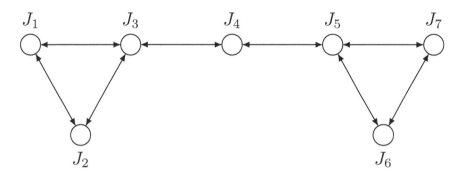

FIGURE 4.1
Example of a primal graph of an urban traffic network

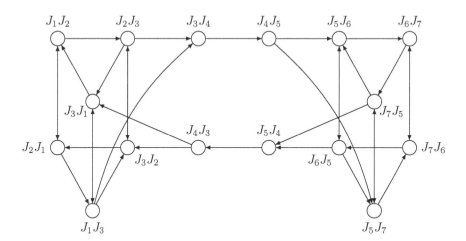

FIGURE 4.2
The dual graph corresponding to the graph shown in Figure 4.1

Markov chain is constructed in the following way. Each intersection in the road network is a vertex in the graph, and there is an edge between two vertices if there is a road segment that connects the two corresponding intersections.

An example of such a graph can be found in Figure 4.1, where junctions $J_1, J_2, \ldots, J_7$ are connected by road segments. For instance, the road segment $J_1 J_2$ is the road that allows a vehicle to drive from J_1 to J_2, and is different from $J_2 J_1$ which goes from J_2 to J_1. This graph is called *primal graph*. On the other hand, in the *dual graph* the nodes are the edges of the primal graph, (i.e., the road segments of the traffic network), and there is an edge between two nodes if it is possible to make a direct transition between the corresponding road segments. The dual graph corresponding to Figure 4.1 can be found in Figure 4.2. It can be noted that dual graphs carry more information than the corresponding primal graphs. For instance, we can see from Figure 4.2 that cars are not allowed to perform U-turns at junction J_4, while the same information cannot be recovered from the primal graph in Figure 4.1. The weights for the edges in the dual graph are given by the turning probabilities. In the following, we are interested in the Markov chain that corresponds to the dual graph.

Remark 4.1. The diagonal elements of the transition matrix $\mathbf{P}$ are related to the probability to remaining in the same state (i.e., the same road). This quantity is proportional to the average time that vehicles spend along a given road which depends, among others, on the length of the road, on the average congestion, and on the speed limits on the road. If we let $\bar{t}_i > 1$ denote the average travel time required to travel along the ith road, then the diagonal entry corresponding to the ith road can be computed as $\mathbf{P}_{ii} = (\bar{t}_i - 1)/\bar{t}_i$ (see the Appendix of [40]). Note that $\bar{t}_i$ needs to be greater than one, as values smaller than one would imply that a vehicle could be able to drive along more than one road in less than a time step of the Markov chain (also, $\bar{t}_i$ smaller than one would cause negative entries in the matrix $\mathbf{P}$). A more detailed discussion of the unit of measurement, or the step size in the computation of the diagonal entries can be found in [162]. Finally, note that the presence of at least one positive diagonal element in matrix $\mathbf{P}$, together with the assumption of $\mathbf{P}$ being an irreducible matrix, guarantees that $\mathbf{P}$ is primitive as well (see [114]).

4.2.3 Benefits of Using Markov Chain to Model Mobility Dynamics

Markov chains appear to be particularly suitable for traffic applications for a number of reasons:

- Microscopic behavior is embedded into the chain through aggregation. For this purpose, data can be easily measured (e.g., travel times and turning probabilities);

- The suitability of Markov chains for big-data applications is discussed, for example, in the context of Google's PageRank algorithm ([114]). Accordingly, well-established and robust algorithms are already available to

TABLE 4.1

Interpretation of MC quantities in the corresponding MC

Quantity/Markov Chain	Traffic MC	EV Traffic MC
Perron eigenvector	Vehicular density in the network	Energy consumption density
MFPT	Average time required to go from i to j	Average energy required to go from i to j
Kemeny constant	Average time required for a random trip	Average energy required for a random trip
Second eigenvector	Neighbourhood where vehicles spend most of their time	Neighbourhoods where EVs consume most energy from their batteries

handle datasets of the size of millions, and more, webpages. This property appears particularly convenient to model urban networks composed by a large number of roads. As an example, the model described above has been validated over real data of North Jutland in Denmark (where the primal road network has 17956 vertices and 39372 edges) and of the city of Beijing, see [135], [133] and [199]. Note also that the suitability of Markov chains to capture and model complex dynamical systems was initially discussed in [67] and in the context of urban networks in [161];

- The proposed model opens the way to multi-variate and derivative models. Namely, while the basic Markov chain describes the probability of making a transition from one road segment to another in one time step, it is possible to easily switch to a unity of energy instead of time. As a consequence, the new Markov chain describes the energy consumption of EVs in an urban network, rather than the time spent in the same network;

- Many of the properties of the Markov chain have a nice straightforward interpretation in the mobility counterpart. For instance, Table 4.1 gives the interpretation of Markov chain quantities in the conventional mobility case, and in the EV case considered here.

The next section illustrates in a greater detail how the traffic model for EVs can be derived from the conventional model described in Section 4.2.2.

4.2.4 Energy Consumption in a Markov Chain Traffic Model of EVs

In the basic Markov chain transition matrix $\mathbf{P}$, the step unit is a unit of time, i.e., a vehicle goes from one state to another (or possible the same) state with a certain probability after a unit of time. In some cases however it may be desirable to change the step unit from a unit of time to something different (e.g., a unit of pollution as in [41]). In this context, we are interested in considering the energy expended by EVs in traveling along one road, rather than the time required to do so. For this purpose, the matrix $\mathbf{P}$ can be transformed into another Markov chain transition matrix $\mathbf{Q}$ as

$$\mathbf{Q} = (\mathbf{I} - \mathbf{D})\,\mathbf{P} + \mathbf{D}, \tag{4.2}$$

where $\mathbf{D} = \mathrm{diag}(\tilde{w}_1, \ldots, \tilde{w}_n)$ is a diagonal matrix of weights $\tilde{w}_i, i = 1, \ldots, n$, n is the number of road segments in the network of interest, and $\mathbf{I}$ is the identity matrix of appropriate dimensions. The weights $\tilde{w}_i, i = 1, \ldots, n$ can be chosen as

$$\tilde{w}_i = \frac{w_i - \alpha}{w_i}, \tag{4.3}$$

where $w_i, i = 1, \ldots, n$ correspond to the average cost associated with traveling through the ith road segment (e.g., in terms of battery consumption) and $0 < \alpha \le w_i$ is the cost associated with one step of the Markov chain, and will be denoted as the step size. In the Markov chain EV model, the unit of the Markov chain is a unit of energy, associated with the average energy consumption from the battery in the EVs. For this purpose, we adopt in the following a simple model that is taken from [162] and represents an extension of the simple *constant speed model* inspired by [129].

We shall assume a driving pattern where vehicles are stationary at the beginning of each road segment (i.e., speed equal to zero) and accelerate at a constant rate a_1 until they reach the cruising speed, v_{cruise}, and then decelerate at a constant rate a_2 to reach the end of the road segment with zero velocity. The energy can be calculated as the sum of the energies in the acceleration phase, the cruising phase and the deceleration phase. In particular, the following forces affect the vehicle:

- $F_{\text{acc}} = ma$ is the force acting on the vehicle while it accelerates to, or decelerates from, the cruising speed

- $F_{\text{rol}} = \mu_{\text{rol}} mg$ is the force needed to overcome the rolling resistance

- $F_{\text{ad}} = \frac{1}{2}\rho A C_d v^2$ is the aerodynamic drag force

- $F_{\text{slope}} = mg\sin(\phi)$ is the hill climbing force,

where a is the vehicle's acceleration, A is its frontal area, m its mass and v its speed, $\mu_{\text{rol}}, \rho, C_d$ and g are constants and ϕ is the inclination of the road segment. For our calculations we assume a car weight of 1235 kg, gravitational acceleration of $g = 9.81\,\text{m/s}^2$. Further reasonable parameter choices for a medium sized EV are: $\rho = 1.2\,\text{kg} \cdot \text{m}^3$, $\mu_{\text{rol}} = 0.01$, $C_d = 0.35$, $A = 1.6\,\text{m}^2$ [129]. We shall assume that the slope along a road segment is constant. In general (given an accurate velocity profile) we can calculate the energy expended as the integral of the force over the path. As discussed above we split this in three parts.

$$W = \underbrace{\int_0^{x_1} (F_{\text{acc}} + F_{\text{rol}} + F_{\text{ad}} + F_{\text{slope}})\, dx}_{W_1} +$$

$$+ \underbrace{\int_{x_1}^{x_2} (F_{\text{rol}} + F_{\text{ad}} + F_{\text{slope}})\, dx}_{W_2} +$$

$$+ \underbrace{\int_{x_2}^{x_3} (F_{\text{acc}} + F_{\text{rol}} + F_{\text{ad}} + F_{\text{slope}})\, dx}_{W_3},$$

where x_1 is the distance after which the cruising speed is reached and x_2 is the distance after which the deceleration process is started and x_3 corresponds to the length of the whole road segment. If we regard speed and distance as functions of time t and denote them $v(t)$ and $s(t)$ respectively, then speed, time, and distance are related by

$$v(t) = v(0) + at$$

and

$$s(t) = s(0) + v(0)t + \frac{1}{2}at^2,$$

where $a \in \mathbb{R}$ is the constant acceleration. We also assume a constant acceleration and deceleration of $a_1 = -a_2 = 3\,\text{m/s}^2$. If the length of the road, x_3, is large enough, we obtain

$$x_1 = \frac{1}{2}a_1 t_{\text{acc}}^2 = \frac{1}{2}a_1 \left(\frac{v_{\text{cruise}}}{a_1}\right)^2 = \frac{1}{2}\frac{v_{\text{cruise}}^2}{a_1}$$

and

$$x_2 = x_3 + \frac{1}{2}\frac{v_{\text{cruise}}^2}{a_2},$$

where t_{acc} is the time it takes to accelerate the vehicle to cruising speed at rate a_1 and v_{cruise} is the cruising speed. Thus

$$W_1 = \int_0^{x_1} \left(ma_1 + \mu_{\text{rol}}mg + \frac{1}{2}\rho A C_d v^2 + mg\sin(\phi)\right) dx$$

$$= \frac{1}{2}mv_{\text{cruise}}^2 + \frac{1}{2}\frac{v_{\text{cruise}}^2}{a_1}mg\left(\mu_{\text{rol}} + \sin(\phi)\right) + \frac{1}{8}\rho A C_d \frac{v_{\text{cruise}}^4}{a_1}.$$

For the second integral we obtain

$$W_2 = \int_{x_1}^{x_2} \left(\mu_{\text{rol}}mg + \frac{1}{2}\rho A C_d v^2 + mg\sin(\phi)\right) dx$$

$$= \left(x_3 + \frac{1}{2}\frac{v_{\text{cruise}}^2}{a_2} - \frac{1}{2}\frac{v_{\text{cruise}}^2}{a_1}\right)\left(mg\left(\mu_{\text{rol}} + \sin(\phi)\right) + \frac{1}{2}\rho A C_d v_{\text{cruise}}^2\right),$$

and for the third integral

$$W_3 = \int_{x_2}^{x_3} \left(ma_2 + \mu_{\text{rol}} mg + \frac{1}{2}\rho A C_d v^2 + mg\sin(\phi) \right) dx$$

$$= \left(ma_2 + mg\left(\mu_{\text{rol}} + \sin(\phi)\right) \right) \left(-\frac{1}{2}\frac{v_{\text{cruise}}^2}{a_2} \right) - \frac{1}{8}\rho A C_d \frac{v_{\text{cruise}}^4}{a_2}.$$

We further assume that the losses along the drive train are a constant 15% and during deceleration 50% of the energy can be saved by regenerative braking. An additional and highly important factor in the consumption of battery load is on-board equipment such as heating, light, air-conditioner, radio and many others that draw power at constant rate over time. Their demand is much harder to model as it depends on the individual driver, but their aggregate effect cannot be neglected as it will be further elaborated in Chapter 15. We now have a way to approximate energy requirements for traversing given road segments. Because of the regenerative braking it is possible that the energy requirement takes a negative value, meaning that a vehicle gains energy by traversing that road segment. In the next section we deal with the question, as to how our framework extends to the case where we have negative weights on some road segments.

Remark 4.2. The previous model allows one to construct the Markov chain transition matrix using only measured traffic data (e.g., average speeds) and some static data (e.g., slopes of the roads). However, it is possible to directly use the proposed Markov chain approach also by directly measuring average battery consumption data, if available.

4.2.5 Dealing with Negative Entries

In Section 4.2.4, in (4.2) positive weights w_i were used to build a diagonal matrix $\mathbf{D}$ that in turn transformed the basic Markov chain $\mathbf{P}$ into a new Markov chain $\mathbf{Q}$. The new transition matrix $\mathbf{Q}$ was characterized by a different step unit than the original transition matrix $\mathbf{P}$. However, in the case of EVs, the mechanism of regenerative braking can cause non-positive entries in some of the weights w_i. In order to apply the previously developed theory for the conversion mechanism, we shall make the assumption that the energy required to travel along a road i cannot be exactly zero, otherwise the corresponding weight w_i would be equal to 0 and in turn this would disallow using (4.2) to calculate the diagonal matrix $\mathbf{D}$. In addition, note that negative weights imply that $\mathbf{Q}$ may not necessarily be a transition matrix of a Markov chain anymore, it might not be stochastic (since it might not be non-negative and some values may be greater than 1), so standard methods to analyze Markov chains would not apply anymore to matrix $\mathbf{Q}$. In particular, with negative weights, the matrix $\mathbf{Q}$ and its eigenvectors do not seem to have a straightforward interpretation as in the case of all positive weights.

To be able to use the theory of Markov chains previously described, we define an intermediate Markov chain whose step unit is a unit of energy exchanged between the vehicle and the road network, regardless of whether such a unit of energy was spent or gained (thanks to regenerative braking). To do this, let $\mathbf{W} = \mathrm{diag}\,(w_1, w_2, \ldots, w_n)$, and let $|\mathbf{W}|$ be the diagonal matrix that contains the absolute values of all weights, $|\mathbf{W}| = \mathrm{diag}\,(|w_1|, \ldots, |w_n|)$. Accordingly we define $\widetilde{\mathbf{D}} = \mathbf{I} - \alpha|\mathbf{W}|^{-1}$, where now $0 < \alpha \le \min_i |w_i|$. We then obtain the transition matrix of the intermediate chain from

$$\widetilde{\mathbf{Q}} = (\mathbf{I} - \widetilde{\mathbf{D}})\mathbf{P} + \widetilde{\mathbf{D}}. \tag{4.4}$$

We are interested in keeping track of the sign of energy exchange (i.e., to compute the actual energy required to travel a given route). We assume that transitions between streets occur at the end of the energy step, and that the gain or loss of energy while driving along road segment i is independent of the choice of the next road segment j. With these assumptions in place, we introduce the notation σ_i to indicate the sign of the change in energy transferred from the vehicle to the network. That is, $\sigma_i = 1$ if the vehicle loses energy driving along road i, and $\sigma_i = -1$ if the vehicle gains energy driving along road i.

As the quantity of interest is energy instead of time, we use the term mean first passage energy (MFPE) instead of mean first passage time in this context. We use the intermediate Markov chain matrix $\widetilde{\mathbf{Q}}$ to calculate a generalized version of MFPE (generalized to include possible negative values), following and extending the approach of [78]: For $j \ne i$, to calculate m_{ij}, the MFPE from i to j, we observe that in going from i to j we make a direct transition with probability $\widetilde{q}_{ij}$ and spend σ_i units of energy. With probability $\widetilde{q}_{ik}$ we make a transition to $k \ne j$ where we again spend σ_i units of energy to get to k, in addition the expected energy required to get from k to j is equal to m_{kj}. Thus for $j \ne i$

$$m_{ij} = \widetilde{q}_{ij}\sigma_i + \sum_{k \ne j}\widetilde{q}_{ik}(m_{kj} + \sigma_i) = \sum_{k \ne j}\widetilde{q}_{ik}m_{kj} + \sum_{k=1}^{n}\widetilde{q}_{ik}\sigma_i = \sum_{k \ne j}\widetilde{q}_{ik}m_{kj} + \sigma_i.$$

For any fixed $j = 1 \ldots, n$ we can write this in vector form

$$m_{(j)} = \widetilde{\mathbf{Q}}_{(j)}m_{(j)} + \sigma_{(j)}, \tag{4.5}$$

where $m_{(j)} = \left(m_{1j}, m_{2j}, \ldots, m_{(j-1)j}, m_{(j+1)j}, \ldots, m_{nj}\right)^{\top}$ and equivalently $\sigma_{(j)}$ is the vector of all σ_i for $i \ne j$ and $\widetilde{\mathbf{Q}}_{(j)}$ is obtained from the matrix $\widetilde{\mathbf{Q}}$ by eliminating the jth row and the jth column. Now we can calculate m_{ij} for all $i \ne j$ using the formula

$$m_{(j)} = (\mathbf{I} - \widetilde{\mathbf{Q}}_{(j)})^{-1}\sigma_{(j)}, \tag{4.6}$$

where we use the fact that $(\mathbf{I} - \widetilde{\mathbf{Q}})$ is a singular, irreducible M-matrix and according to [18, Theorem 4.16] each of its proper principal submatrices is invertible.

Remark 4.3. Note that (4.6) is identical to a standard formula for calculating mean first passage times in the case of positive weights only. As we have shown it still works if we have negative weights in the chain.

As in the case of positive weights we are interested in the Kemeny constant as a global efficiency measure, but if we introduce negative weights in our graph, then (4.1) is no longer independent of the starting node i. However, we can generalize the notion of the Kemeny constant in the following way. Let π be the Perron eigenvector of $\tilde{\mathbf{Q}}$ and let

$$K = \sum_{i=1,\ldots,n} \pi_i K_i = \sum_{i=1,\ldots,n} \pi_i \sum_{j \neq i} \pi_j m_{ij}. \tag{4.7}$$

Then K coincides with the Kemeny constant as defined in (4.1) in the case where all weights are positive.

4.3 Sample Applications

One of the main advantages of the derived model is that it can be used to predict traffic flows and to facilitate the taking of pre-emptive measures to mitigate congestion or to balance traffic in general. In Chapter 5 we shall see in detail how the model can be used to derive alternative non-conventional routing advisors. In this section we shall now illustrate the use of the Markov model to balance traffic in a decentralized manner.

4.3.1 Traffic Load Control

Being able to control the stationary distribution of a road traffic network opens up a wide range of possibilities and different applications. In many cases, it is possible to plan optimal target stationary distributions and try to drive the system towards them. For example, using the congestion chain it may be possible to equalize the traveling time on alternative routes to distribute the load within the network. Alternatively road network designers may be able to unburden some road segments in order to facilitate maintenance activities. In the context of EVs, the control of the entries of the Perron eigenvector corresponds to control where energy is consumed within a network of EVs. Thus, from a network designing point of view, it may be interesting to be able to match high energy consumption with free charging point capacity. This can be used to balance the energy demand at charging stations, and thus to potentially reduce the queues of vehicles requiring charging.

TABLE 4.2
Speed limits (in km/h) in the uncontrolled case, in the unrealistic optimal solution, and in the realistic balanced case

Road Segment	J_1J_2	J_1J_3	J_2J_1	J_2J_3	J_3J_1	J_3J_2	J_3J_4	J_4J_3
Uncontrolled	50	50	50	50	50	50	50	50
Optimal (Unrealistic)	15	44	15	29	44	29	117	122
Realistic Solution	30	50	30	40	50	40	100	100

Road Segment	J_4J_5	J_5J_4	J_5J_6	J_5J_7	J_6J_5	J_6J_7	J_7J_5	J_7J_6
Uncontrolled	50	50	50	50	50	50	50	50
Optimal (Unrealistic)	117	117	45	21	42	10	23	8
Realistic Solution	100	100	50	40	50	30	40	30

4.3.1.1 Theoretical Approach

A simple strategy to regulate the Perron eigenvector of the chain is to influence the diagonal entries of the transition matrix via diagonal scaling. In practice, the scaling can be determined by a host of factors, some of which can be controlled by the network designer. One of these is the speed limit. Recall that if the left Perron eigenvector of $\mathbf{P}$ is $x^\top$ and we wish to achieve through feedback a target left eigenvector $z^\top$ of $\mathbf{Q}$ (as from (4.2)) for some positive vector $z = (z_1, \ldots, z_n)^\top$ we set $w_i = \frac{z_i}{x_i}$ according to Lemma 1 in [162].

A useful application of the Perron vector control is now illustrated through an example which exploits again the road network of Figure 4.2. The dotted line in Figure 4.3 depicts the nominal density of cars in the case of uniform speed limits set to 50 km/h. Let us assume that the road engineer is interested in manipulating speed limits in order to achieve a uniform density of cars along all road segments (traffic balancing). Lemma 1 in [162] can be used to compute the "optimal" weights, and predict the "optimal" speed limits accordingly. We validate our approach using SUMO (Simulation of Urban MOBility), a popular realistic mobility simulator, described in more detail in Chapter 17. The dotted line in Figure 4.3 shows that cars can be balanced indeed, although such "optimal" speed limits are unrealistic, as shown in Table 4.2, second line. A better trade-off is shown with solid line in Figure 4.3, where reasonable speed limits are identified instead (e.g., they are all multiples of 10), as reported in the third line of Table 4.2.

Remark 4.4. The proposed application assumes that drivers will not change their routes as a reaction to the new imposed speed limits; we also varied speed limits under the implicit assumption that this would correspond to proportionally adjusting average travel speeds. Despite these assumptions, Figure 4.3 shows that the road network traffic, as observed from SUMO simulations, is in accordance with the theory.

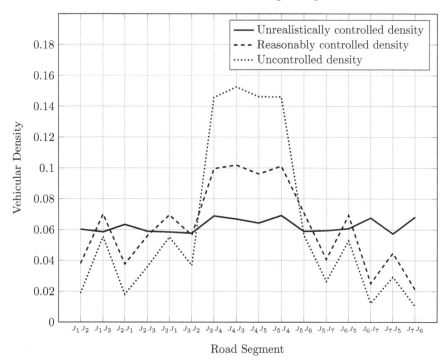

FIGURE 4.3
Comparison of traffic density as a function of speed limits. Uniform speed
limits lead to the unbalanced solution shown with a dotted line. A very good
balance, in solid line, can be achieved using the unrealistic speed limits reported in Table 4.2, second line. A trade-off solution, shown with a dashed
line, is obtained by using realistic speed limits as shown in Table 4.2, third
line

4.3.1.2   Decentralized Traffic Load Control

In the previous section we controlled the Perron vector by choosing appropriate weights in the graph. However, note that while it is known that different
speed limits or different traffic light sequencing affect the weights, it is very
hard to find a precise mathematical relation between them. Thus, it might
be more appropriate not to change speed limits (or traffic light sequencing)
in a single action, but rather to finely tune them until the desired effects are
achieved. In this section we shall use an algorithm from [168] to reach our
goal. In particular, let us assume that a large road network consists of two
main components that are connected by four bridging streets as depicted in
Figure 4.4. We assume that each road segment is able to measure the density
of cars traveling on that road, and to communicate this information to the
neighboring roads.

As illustrated in a different context in [168], our objective here is to determine speed limits that equalize the load across certain road segments. To this end, let $\mathcal{I}$ be the set of road segments of interest, let us discretize time in steps $k = 1, 2, \ldots$ and let speed limits $v_i(0)$ at time $k = 0$ be given for all $i \in \mathcal{I}$. Let us denote by $\theta_i(k)$ the density of energy dissipated on each road segment at the kth time step. Then we use the following iterative equation to update the speed limits

$$v_i(k + 1) = v_i(k) + \eta \sum_j (\theta_j(k) - \theta_i(k)), \tag{4.8}$$

where η is a positive parameter and the sum is taken over all road segments $j = 1, \ldots, n$ that road i can communicate with. The implicit consensus is then conducted by alternating the following two steps:

1. Determine densities of interest (e.g., energy dissipation) at time k.

2. Update speed limits according to (4.8).

We have used SUMO simulation runs and calculated the stationary distribution of the energy Markov chain. We used the entries of the stationary distribution to compute the density of each road segment. We then performed an iterative update of the speed limits according to (4.8). In Figure 4.5 we give the relevant entries of the stationary distribution as a function of the number of simulation steps. In this context we adopted the implicit consensus algorithm in Equation (4.8) without proving that it achieves the balancing of the stationary vector. Full details of the algorithm can be found in [168].

4.4 Concluding Remarks

In this chapter we have reviewed a Markov chain based model to provide a macroscopic description of energy consumption of EVs in a road network. Classic Markov chain theory was generalized to account for some negative entries in the transition matrices, required to model regenerative braking effects. We have also provided some sample applications of the described model, while other applications like routing will be further described in Chapter 5. However, it is our opinion that the same framework may be further extended and other EV related applications may be developed as well; for instance the proposed model may be used to plan the optimal positioning of charging points (i.e., identifying the optimal nodes in the primal network). Another interesting line of research may be the enhancement of the graph to further include other transportation networks, to integrate the reduced range of EVs with other means of transportation (e.g., bikes). This would eventually lead to so-called multi-modal transportation networks, along the lines of [57].

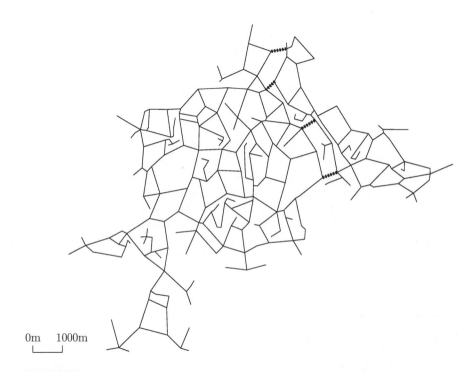

0m   1000m

FIGURE 4.4
Scenario of a big city and a suburb with 4 connecting streets, where we try to
equalize the amount of energy expended on each street out of the suburb

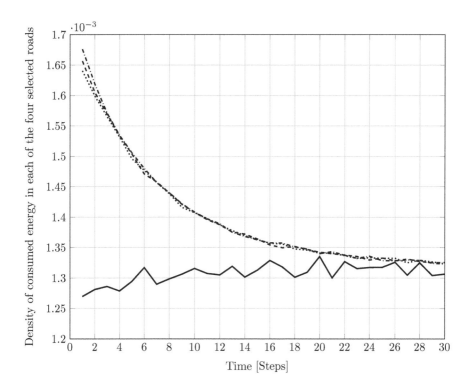

FIGURE 4.5

Convergence results for the suburb scenario. The vertical axis shows the corresponding entries of the Perron vector. Each time step corresponds to one simulation instance

5

Routing Algorithms for EVs

5.1   Introduction

Most drivers of conventional vehicles are already familiar with routing services. Roughly speaking, a routing adviser computes the optimal route that connects an origin point with a destination point. Typically, the user has a choice on the optimality criterion that is used to compute the optimal path, where the most popular choices are the *shortest* path (e.g., in km), the *minimum time* path, the *most economic* path (e.g., where toll charges are minimized) and the *minimum fuel* path (i.e., where fuel consumption is taken into explicit account). Shortest path search algorithms are also well established in the literature, where classic approaches include Bellman's routing solutions [16], Dijkstra's algorithm [52] and dynamic programming algorithms [19]. Sometimes, the optimality of the solution is relaxed in order to obtain routing suggestions in a reasonable amount of time, especially in the case of very large road networks.

Routing algorithms are rather more specialized when applied to electric vehicles. Here, the refueling need of a vehicle is a primary concern due to the limited battery capacity and the relatively long time required for vehicle charging. Such concerns have been investigated by many researchers from the scientific community and a number of different approaches have been proposed. More specifically, it is possible to distinguish three particular categories of routing problems that are of significant interest in the case of electric vehicles:

1. **Route planning (i.e., minimum energy routing):** In this case the topography of the road network, together with (predicted or measured) traffic conditions are used to calculate the minimum energy route to a destination; see for example [138]. Such routing is not only beneficial in terms of energy consumption, but also in other ways as energy efficient routes also increase the battery lifetime due to the reduced number of battery charging cycles [98]. An important aspect of such route computations is the estimation of the energy recuperation (regenerative braking) coefficient, both in route calculation and in the on-board energy optimization [98]. In this context, the elevation profile of the roads

composing the candidate routes has a strong impact on the identification of the optimal route. Other authors have proposed the use of real-world driving patterns extracted from existing databases to pre-compute optimal routing suggestions in terms of energy consumption [115]. The advantage of such an approach is that energy consumption can be evaluated not only as a function of static parameters (e.g., the altitude values), but also as a function of (average) time-stamped values of the velocity. Note that this class of routing algorithms considers "single" vehicles in the sense that the optimal choice is not affected by what other vehicles do in the road network.

2. **Planning of both the route and charging events:** Logistics systems are seen as one of the most attractive fields of application for EVs; in particular due to the lower costs for charging in comparison to those for refueling [200]. In this case, the Electric Vehicle Routing Problem (EVRP) consists of planning the most convenient order in which a set of customers should be served, given the battery constraints, the availability of charging points, and the constraints on the time window in which customers are supposed to be served [5]. This problem is known to be an NP-hard problem, as it is a natural extension of capacitated vehicle routing problems [5]. During the operation time, it is expected that a single vehicle might possibly have to charge several times in order to serve all the customers.

3. **Multi-agent route and charging planning:** In this class of problems, charging events are now explicitly taken into account when recommending optimal routes. In particular, should the battery capacity of a vehicle not be enough to reach the final destination, then the recommended routes will be shaped in order to include a charging point along the path. In this case, it is of paramount importance also to consider the possibility that other vehicles might use the same charging stations, as this will affect the time required for charging and in turn, the time required for the whole journey. Accordingly, the EV routing problem now becomes finding an energy optimal route with appropriate charging stops, and minimal idle waiting time at charging stations [8]. For instance in [187] a Vehicle to Vehicle (V2V) communication protocol is proposed to realize the context-awareness aspect of route planning, while in [8] a centralized solution is explored. Note that in this context both competitive and cooperative approaches can be designed, depending on whether vehicles are expected to compete, or are incentivized to cooperate, to find an available charging point. Note that in this class of routing problems the effect of feedback should not be neglected (i.e., the impact of routing advice on the evolution of traffic flows).

Given this basic background, this chapter is intended to give a very brief qualitative introduction on the topic of routing. As there are many papers on this subject, our intention is to mention some of the approaches that are available in the literature and to point the interested reader to the relevant results.

5.2 Examples of Selfish Routing for EVs

In what follows, a vehicle plans its optimal route without taking into account the actions of other vehicles. For this reason, such algorithms are sometimes called selfish routing algorithms because a macroscopic view of the whole population of vehicles is missing, or opportunistic routing because a vehicle chooses its optimal route without considering the impact of its choice upon the whole population of vehicles or the road network. We shall also present applications that exploit the Markov chain framework illustrated in Chapter 4.

Minimum Energy Routing: As already mentioned, a very simple, yet important, routing strategy for electric vehicles is obtained by minimizing the distance to the destination not in terms of travel time or actual distance, but instead in terms of energy charge needed to finish the journey. Such a solution can be developed using the Markov chain model described in Chapter 4, by using the weights calculated as in Section 4.2.4 in combination with a classic graph search algorithm (e.g., Dijkstra's algorithm [52]).

For instance, let us consider the road network depicted in Figure 5.1 as a case study for minimum energy routing. There are three routes connecting J_2 to J_3. Assume that route (a) is 1.8 km long and is traveled at an average speed of 50 km/h, route (b) is 1 km long and traveled at 80 km/h and route (c) is 1.4 km long and traveled at 80 km/h. Further, assume that route (b) is not flat, but rises up a small hill at its center at a constant slope of 5% in both directions. Table 5.1 reports the energy required to travel each route calculated with two different power levels needed by auxiliary systems in the vehicle. According to [61] the power demand of accessory systems in an electric vehicle varies hugely depending on weather; e.g., the power may vary from a minimum of 500 W to a maximum of 3500 W, for instance in winter, when the heating is running at full power. When traveling from J_1 to J_4, the driver has a choice to take one of three routes (a),(b) or (c). The shortest route between J_2 and J_3 is route (b), but we can see in Table 5.1 that all other (longer) routes perform better when minimizing the required energy for traveling. This is due to the topology of route (b) and the high energy costs for electric vehicles ascending hills. Additionally, we can see that route (a) is more energy efficient than route (c) if the auxiliary power demand is low,

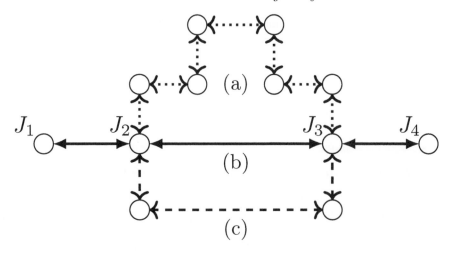

FIGURE 5.1
There are three possible paths to go from J_2 to J_3. Different routing strategies based on minimum distance or minimum energy consumption suggest different paths, also depending on the different environmental conditions (e.g., air conditioning switched on or off).

TABLE 5.1
Required energy (in kWs) to travel from J_2 to J_3 in the road network depicted in Figure 5.1

Auxiliary Power Demand	Low (500 W)	High (3500 W)
Route (a)	535	924
Route (b)	915	1050
Route (c)	695	884

while this relationship is reversed if the auxiliary power demand is high. This implies that, especially in more complex routing tasks, an estimate of the power demand and its changes over time may have an impact on the choice of the most efficient route.

An important conclusion from this simple example is that it is possible to assist users in making energy conscious route choices by providing energy road maps, i.e., road maps in which the displayed distance corresponds to energy consumption instead of travel distance.

Congestion Avoidance Routing: In order to minimize the risk of traffic accidents and congestion along the vehicle's path, in some situations it may be advisable to avoid the roads that most people use. Such popular roads could for example be close to shopping areas or train stations. The basic Markov chain presented in Chapter 4 describes how often roads are taken, while ignoring travel times or energy consumption. Thus, it carries information about popularity and the stationary distribution can be used to find the minimum popularity route, again using conventional graph search algorithms.

Mixed Minimum Energy/Congestion Avoidance Routing: Both minimum energy and minimum popularity routing may be too focused on one aspect of traveling. To circumvent this, it is possible to use a weighted sum of the above two quantities to find an optimal path, in terms of an appropriate trade-off of the two aspects. In this sum we have an additional tuning parameter that can be set according to how important energy and popularity are considered to be. In this context, it is possible to use the Markov chain presented in Chapter 4, which contains information about both energy consumption and popularity and use its stationary distribution as weights for the graph search.

Energy Optimal Safe Routing: This algorithm was first proposed in [42] in the case of conventional ICE vehicles, and had the objective to take into account whether a driver was familiar or not with the territory to be driven. In fact, it makes sense to take into account the probability that the driver makes a mistake and does not follow the exact route recommended by the routing adviser (e.g., a car navigator). In this case, a re-routing suggestion is required to send back the driver along the correct track to reach the desired destination. Clearly, making such mistakes causes an inconvenience to the driver, and in general re-routing has an impact on the overall objective function (e.g., an increased travel time or an increased fuel consumption).

In the specific case of EVs, the consequences of mistakes might be even more critical, as they might give rise to an unexpected increase of energy consumption. Accordingly, the safe-routing algorithm takes into account the familiarity of the driver with the territory to be driven, and recommends the path where the probability of making a mistake is the smallest, or where the consequences are not too critical (e.g., even when taking the wrong path, it is still possible to find an EV charging point nearby). The routing algorithm coincides with a conventional algorithm (e.g., a minimum-time routing algorithm) in the case that the driver confirms that he or she is actually familiar with the territory.

The following simple example from [42] briefly summarizes the main philosophy underlying the safe-routing algorithm. Let us consider the simple road graph shown in Figure 5.2. According to the model outlined in Chapter 4 the first step consists in transforming the primal network shown in Figure 5.2 into the dual one, where nodes correspond to roads. Let us denote the roads in accordance with junctions they connect (i.e., the road from junction J_2 to J_3 is named J_2J_3). For simplicity, we assume that travel times are proportional to road lengths, and are normalized so that road J_2J_3 takes for instance one unit of time, while J_2J_4 requires $\sqrt{2}$ time units. The diagonal entries of the

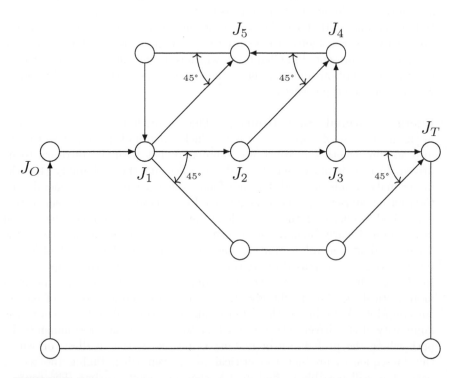

FIGURE 5.2
Example of a primal network. The objective is to compute the best path to go from the origin node J_O to the target node J_T.

TABLE 5.2

Transition matrix $\mathbf{P}_{RW}$

	$J_O J_1$	$J_1 J_2$	$J_2 J_3$	$J_3 J_T$	$J_1 J_T$	$J_T J_O$	$J_5 J_1$	$J_1 J_5$	$J_4 J_5$	$J_2 J_4$	$J_3 J_4$
$J_O J_1$	0	1/3	0	0	1/3	0	0	1/3	0	0	0
$J_1 J_2$	0	0	1/2	0	0	0	0	0	0	1/2	0
$J_2 J_3$	0	0	0	1/2	0	0	0	0	0	0	1/2
$J_3 J_T$	0	0	0	0	0	1	0	0	0	0	0
$J_1 J_T$	0	0	0	0	0.74	0.26	0	0	0	0	0
$J_T J_O$	0.13	0	0	0	0	0.87	0	0	0	0	0
$J_5 J_1$	0	1/6	0	0	1/6	0	1/2	1/6	0	0	0
$J_1 J_5$	0	0	0	0	0	0	0.71	0.29	0	0	0
$J_4 J_5$	0	0	0	0	0	0	1	0	0	0	0
$J_2 J_4$	0	0	0	0	0	0	0	0	0.71	0.29	0
$J_3 J_4$	0	0	0	0	0	0	0	0	1	0	0

TABLE 5.3

Transition matrix $\mathbf{P}$ corresponding to a perturbed optimal path

	$J_O J_1$	$J_1 J_2$	$J_2 J_3$	$J_3 J_T$	$J_1 J_T$	$J_T J_O$	$J_5 J_1$	$J_1 J_5$	$J_4 J_5$	$J_2 J_4$	$J_3 J_4$
$J_O J_1$	0	$1-\epsilon$	0	0	$\epsilon/2$	0	0	$\epsilon/2$	0	0	0
$J_1 J_2$	0	0	$1-\epsilon$	0	0	0	0	0	0	ϵ	0
$J_2 J_3$	0	0	0	$1-\epsilon$	0	0	0	0	0	0	ϵ
$J_3 J_T$	0	0	0	0	0	1	0	0	0	0	0
$J_1 J_T$	0	0	0	0	0.74	0.26	0	0	0	0	0
$J_T J_O$	0.13	0	0	0	0	0.87	0	0	0	0	0
$J_5 J_1$	0	1/6	0	0	1/6	0	1/2	1/6	0	0	0
$J_1 J_5$	0	0	0	0	0	0	0.71	0.29	0	0	0
$J_4 J_5$	0	0	0	0	0	0	1	0	0	0	0
$J_2 J_4$	0	0	0	0	0	0	0	0	0.71	0.29	0
$J_3 J_4$	0	0	0	0	0	0	0	0	1	0	0

random walk transition matrix $\mathbf{P}_{RW}$ take into account the travel times, while the off-diagonal entries are given the same probability. The entries of the matrix are reported in Table 5.2, where for further convenience the columns and row entries (i.e., roads) are specified.

Dijkstra's algorithm can then be used to compute the minimum time path using the knowledge of travel times, and the optimal path is clearly $J_O J_1 - J_1 J_2 - J_2 J_3 - J_3 J_T$ with an overall travel time of 4 time units, while the second best path (denoted as the alternative path in the following) is $J_O J_1 - J_1 J_T$ which takes $2 + 2\sqrt{2}$ time units. We now perturb the two optimal paths assuming a probability of making mistakes equal to ϵ, so the perturbed transition matrix associated with the minimum time path becomes as in Table 5.3

Obviously the diagonal elements are not modified because they are only related to travel times. Similarly it is possible to perturb the transition matrix associated with the alternative path. The obtained transition matrices are used to compute the corresponding mean first passage time matrices and to compare the expected mean first passage times from the origin road to the destination (e.g., entry $(1, 6)$ of the mean first passage time matrix).

Then, it is possible to observe that for values of ϵ smaller than 8.69%, the minimum time path is indeed the most convenient one; for values of ϵ greater than 8.69%, the alternative path becomes the most convenient one (i.e., it is the safest one); while when ϵ is exactly equal to 8.69%, then the two paths are equally convenient (see [42]).

Therefore, the safe routing path coincides with the minimum time path when $\epsilon = 0$, while it may be different, as in the given example, for larger values of ϵ.

5.3   Collaborative Routing

5.3.1   A Motivating Example

Clearly "selfish routing" only makes sense if very few cars make routing choices. As more and more cars make use of route assist technologies, such an assumption is becoming more and more unrealistic. In particular, one known problem of routing advisers is that if all vehicles that have the same (or close) origin-destination pairs are recommended to use the same route, then local congestion events might occur along the common recommended route. Such a problem may be even more compelling in the case of EVs, as in addition there might be queues at the charging points along the same route. Thus, balancing vehicles along different routes that connect the same origin and the same destination becomes again more convenient in the case of EVs. This problem is now illustrated through a simple example, inspired by the one presented in [163].

Let us assume that a group of vehicles has the same origin and the same destination. We also assume that there are three similar routes from the origin to the destination, that require on average 60, 70 and 80 minutes of travel time. Along each route there is exactly one charging point after 30 minutes of traveling. Let us assume that the initial level of the batteries of the vehicles does not allow them to reach the destination without charging. In the simulation, we assume that the average time for charging is about 15 minutes. In principle, all three routes are feasible for the vehicles. However, if all vehicles follow an opportunistic routing strategy, most likely all of them will take the shortest route (i.e., in terms of travel time). This leads to congested demand for charging along that route. We compare that solution with another one where we assume that a central routing adviser forces vehicle i to take a particular route through some sort of automated booking system with the objective of balancing the charging load. The routing adviser simply associates an EV with the route with (current) minimum load, where the load both refers to the cars currently charging, and those that have already made a reservation at the charging point. As can be seen from Figure 5.3 and Figure 5.4, this algorithm manages to balance both the number of cars along each possible route and the quantity of energy required at each charging station. At the end of the simulation 80 vehicles have completed their journey with an average journey time of about 1 hour and 50 minutes. Just to give an idea of how things would have degenerated with the opportunistic routing strategies, if all vehicles chose the shortest path, which apparently is the best solution, then only 35 vehicles would have finished their journey by the end of the simulation with an average time of more than 5 hours.

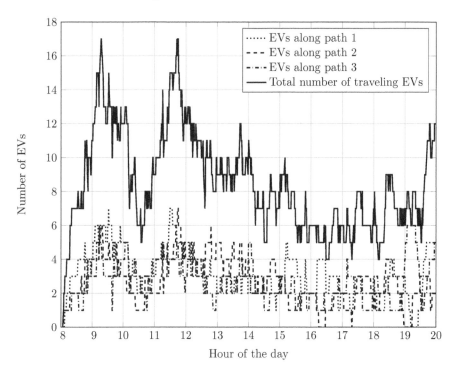

FIGURE 5.3
With a centralized routing adviser, the number of EVs is fairly distributed along the available routes.

5.3.2 Collaborative Routing under Feedback

The previous example showed that centralized routing strategies can be in some cases more efficient than opportunistic routing strategies. In this section we further, briefly, mention recent ideas in collaborative routing solutions.

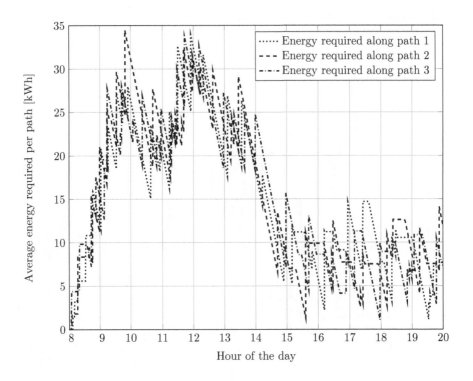

FIGURE 5.4
With a centralized routing adviser, also the energy load can be well balanced
among the three charging points.

The discussion here follows some of the results gathered in [86] and [125]. In particular, the following aspects are relevant when designing a collaborative routing strategy:

- **Flapping instability:** Poorly designed routing systems are prone to flapping. This is a phenomenon where congestion is not alleviated by collaborative routing; rather, traffic congestion is simply shifted from one part of the network to another part, as a consequence of ill-conceived routing signals.

- **Need for closed-loop design:** Routing algorithms are in some cases designed in an open-loop fashion (e.g., historical data are used to predict the level of congestion in the road network). Most cities have elastic requirements necessitating the need for closed-loop routing algorithms where actual traffic information is considered. Note that closed-loop solutions are expected to be more convenient than open-loop strategies, as they also take into account the most recent available traffic data, and use them to influence the choice of the recommended route.

- **Delay induced instabilities:** Routing instabilities can also be induced if congestion information is "old" or, in other words, if there are delays in communicating congestion information to drivers. This is a typical instability induced by delays in a feedback loop.

- **Poor trade-off between utility of the community and utility of individuals:** Routing algorithms can be designed to maximize the benefit to the driver or to the community. Sometimes the needs of both stakeholders are conflicting. For example, to minimize pollution, more polluting vehicles may receive preferential routes if these are shorter.

- **Community constraints:** Most routing algorithms do not take into account network and queueing constraints that arise from societal behavior. Fair access to routes is almost never considered in the context of routing algorithms (e.g., in the example in the previous section, the longest route should not be always recommended to the same driver).

Preliminary collaborative routing strategies have been applied in the context of research projects (e.g., TEAM[1]) with the main objective of balancing the density of vehicles across a network, under a feedback signal. In particular, this can be performed by pursuing a number of different goals: namely, to balance congestion; to balance travel times; to balance battery consumption; or to balance aggregate emissions. A number of use cases have been considered as well, namely, to balance parking demand; to balance traffic load; to balance queues at EV charging points; or to avoid pollution peaks in the close proximity of sensitive areas (e.g., kindergartens, hospitals). In terms of scientific research, the challenges in developing collaborative routing strategies rely on developing

---

[1]http://www.collaborative-team.eu/. Last Accessed July 2017.

algorithms that preserve user privacy (i.e., not to share the information about where single individuals go); guarantee fair access to the resource for all vehicles (i.e., EV charging points; or time-minimum optimal routes); and impose minimal communication with the centralized or distributed routing infrastructure in a "plug and play manner". Different collaborative routing algorithms may be distinguished among those that try to model users' behaviors and anticipate their routing preferences; and those that try to influence and steer the choices of the drivers via penalties and pricing functions to convince them in taking a (collectively) convenient route.

5.4   Concluding Remarks

This chapter gave a brief introduction on the topic of routing, both considering opportunistic and collaborative routing strategies. Especially research related to the second class of routing strategies is still in its nascent stage, and while some preliminary examples have been discussed in Section 5.3.2, this still remains a very active area for research, requiring new ideas and new algorithms.

6

Balancing Charging Loads

6.1 Introduction

In the recent years, we have seen a continuous improvement in the availability of charge points for EVs in many countries. For large numbers of charge points, it becomes interesting to treat the problem of allocating charge points as an optimization problem. For an individual EV, the charging process duration depends on the battery capacity installed on-board (typically from 20 kWh to 85 kWh), the initial SOC, and the type of the charge point. Thus, the charging time can range from 15 minutes up to 10 hours or more, depending on these factors [24]. While such time scales might be acceptable when the vehicle is only charged at night time, things change if the car owner needs charging during a journey to reach the final destination. In such situations, in the presence of queuing, it may take up to an hour until the battery is fully loaded, thus pushing such fast charging stations into the realm of "not acceptable". Such considerations have led to a host of research papers on this topic in recent years [82] [58]. Important examples include [50], where vehicles compute a routing policy that minimizes the expected journey time while considering the *intentions* of other vehicles (i.e., intentions of charging en route); and [30], where a reservation-based charging approach is proposed. Much of this work appears to have been motivated by [87], which initially proposed a distributed algorithm for balancing demand over a network of charge points. Our objective in this chapter is to revisit this work and to describe its main features.

The problem of balancing the charging load across a set of charging stations can be tackled in a number of ways. Classic approaches to this problem include routing individual vehicles in a manner to balance load, or, given average traffic densities, placing the charging stations in a manner to service the expected demand in a balanced fashion. Such charging station positioning problems are already a topical problem in the operations research community [195] and [179]. Notwithstanding the fact that these problems are extremely interesting from a theoretical perspective, they do suffer from a number of practical drawbacks. First, traffic densities are temporal in nature. Consequently, an optimal placement of charging stations for morning traffic might be sub-optimal in evening traffic or may vary over time, thus rendering its usefulness somewhat questionable. Similarly, load balanced routing strategies place severe constraints on drivers by forcing them to take particular routes and by forcing them to particular charging stations. The willingness of drivers to follow recommendations is a significant issue in the design of such systems. With this latter point in mind, another approach might be to let the driver choose a route and then use some sort of admission control to balance the occupancy of charging stations along this route. Such strategies have high data-gathering and communication requirements and again place severe constraints on vehicles and drivers in a way that may make such a system unacceptable to users in practical situations. It is therefore of interest to investigate other balancing solutions. As before, the objective is to balance charging demand across a number of charging stations. We also wish to create a system that adapts to changing traffic densities (flexible infrastructure) and which can be implemented without centralized communication between infrastructure and vehicles. A further objective is to realize the system in a plug-and-play manner; namely, charging stations can enter and leave the system as they wish, without placing a reconfiguration requirement on a centralized infrastructure.

6.2 Stochastic Balancing for Charging

The starting point is an analogy with the mobile cellular network. We view (electric) vehicles as phones and charging stations as access points. The idea is to divide a city area into partition cells and within each cell to associate each vehicle with one charging station. The division of partition cells can be based on graph clustering techniques (e.g., Voronoi diagrams) that make use of electric vehicle density patterns, so that the ratio of EVs to charging stations is roughly constant in each cell. Note that this partition could be easily adapted based on temporal density patterns. The main feature of this kind of algorithm is that drivers are not forced to take selected routes; rather, as they travel from one partition cell to another they are assigned to a charging station in a manner akin to hand-off in mobile phone networks. The partitioning of the city into cells can be realized in many ways and this is not our focus here. Rather, our focus is the car assignment problem within a cell which takes place over faster time scales than the partitioning problem. To achieve this we propose the following stochastic algorithm. Within a partition cell, charging stations broadcast a green signal, indicating their ability to accept a new vehicle, with a frequency that is a decreasing function of the current queue length. Hence, charging stations with more spare capacity advertise their service with a higher frequency. Similarly, we assume that vehicles listen to green signals with a frequency that is a decreasing function of the current level of their battery. If, at a given time, a vehicle senses a green light from a charging station, then the vehicle is assigned to the charging station. In this way, vehicles that are more in need of charge are associated with the charging stations that have the shortest queue in a fully decentralized fashion. The actual implementation of the protocol is described in detail in the following section.

6.3 Basic Algorithm

6.3.1 Charging Stations

Algorithm 6.1 Charging station

$p_{CP}^{(i)} \leftarrow F(N_i)$

Broadcast green light with probability $p_{CP}^{(i)}$

We assume that at every time step the ith charging station communicates its availability to accept a new vehicle with probability $p_{CP}^{(i)}$. To do this it uses Algorithm 6.1, where N_i is number of vehicles currently queuing for charging at the ith charging station, and F is some function mapping the occupancy to $[0, 1]$. We assume that $F(0) = 1$ and as the queue grows F monotonically decreases to 0. As charging stations with a shorter queue are more likely to accept a new car than charging stations with a longer queue, this mechanism is able to balance the number of vehicles associated with each charging station.

6.3.2 Electric Vehicles

Algorithm 6.2 Electric vehicle

$p_{EV}^{(i)} \leftarrow G(e_i)$

Listen for green signals with probability $p_{EV}^{(i)}$

Once the ith electric vehicle has decided to search for a charging station we assume that this vehicle at each time step senses a green signal according to Algorithm 6.2, where G is a function that maps the vehicles current SOC, e_i, to a probability. We require that G is a decreasing function and $G(0) = 1$. Throughout this chapter we use the following way to compute $p_{EV}^{(i)}$

$$p_{EV}^{(i)} = 1 - \frac{e_i}{M}, \tag{6.1}$$

where e_i is the actual level of energy in ith EV, and M is a maximum quantity of energy. The rationale behind Equation (6.1) is that EVs with low charge are more likely to sense a green signal than vehicles with plenty of residual charge, thus giving EVs priority according to their requirements.

Remark 6.1. There are many possibilities to refine the basic algorithm. For example, instead of balancing the *number* of vehicles, it is also possible to balance the *energies* requested at each charging station. It is also possible to take into account other factors in the computation of $p_{EV}^{(i)}$, for example the expected energy required to reach a charging station or the distance to it, the expected time to reach the destination and the user needs, such as the accepted detours, location preferences, etc. It may however be justified to ignore these factors, if one assumes that the partition cells are small enough so that each station in a cell is equally acceptable to the user.

6.3.3 Protocol Implementation

Each second, during which a charging station gives a green light and an EV senses such a broadcast, the EV is assigned to the charging station. Note that contrary to most approaches (see for instance [69]) the proposed approach is fully distributed. It is worth mentioning at this point that a fully centralized solution to this problem could easily be implemented. Indeed, such solutions are in many cases preferable to distributed ones. We focus on distributed solutions as they can be very attractive for this application for a number of reasons: (i) such algorithms are usually robust to possible failures, including attacks and manipulations, in a way that centralized solutions are not; (ii) in addition, the approach could easily be extended by introducing a reputation system to identify misbehaving vehicles or charging stations; (iii) distributed self-organizing optimization algorithms are very suitable to handle scenarios which are highly stochastic in nature (e.g., see for example the Transmission Control Protocol (TCP) in communication networks or [145] in the context of vehicular mobility); (iv) the decentralized approach we describe has certain privacy and security benefits, as there is no need for an entity, which knows about the locations of all individual vehicles. On the contrary, decisions can be made in a decentralized manner with only limited information; (v) finally, decentralized solutions facilitate the possibility of implementing plug-and-play policies, where charging stations become available only at particular moments of the day. For example private car parks, office blocks, and universities may make their charging stations available during off-peak times. The latter point needs to be emphasized and will be discussed later through an example. In a centralized solution the adding of charging stations requires updates to the centralized structure, whereas a self-organizing structure leads to a truly plug-and-play system.

Remark 6.2. The implementation of the proposed protocol in a probabilistic framework rather than through a more straightforward deterministic approach is preferable for the following reason. A single charging station cannot know which charging station has the shortest queue unless some information is exchanged among stations (either via a centralized hub, or directly). The proposed approach does not require any direct exchange of information and thus has good scalability properties (with numbers of vehicles and stations).

6.4 Analysis

6.4.1 Quality of Service Analysis: Balancing Behavior

The algorithm described in the previous sections is best analyzed in a queueing theory framework. We now describe a possible approach to do this and give results, under simplifying assumptions, regarding the algorithm for two distinct scenarios. In this section we let the probabilities $p_{CP}^{(i)}$ for Algorithm 6.1 be calculated according to

$$p_{CP}^{(i)} = 10^{-N_i}, \tag{6.2}$$

where N_i is the number of vehicles at the ith charging station. To ease exposition, we make the following assumptions:

(i) There is no delay between a car requesting service and assignment to a charging station.

(ii) After being assigned, the vehicle is instantly added to the charging station's queue.

(iii) We assume that a car is assigned to charging station i with probability $\frac{p_{CP}^{(i)}}{\sum_j p_{CP}^{(j)}}$, where the sum in the denominator is taken over all charging stations in the same partition cell as station i.

(iv) Each car waiting at a charging station is charged with a rate independent of the total number of vehicles awaiting service.

(v) We model the arrival process of cars to the charging stations as a Poisson process and assume that the charging times for each car are exponentially distributed.

Remark 6.3. The first assumption is justified, as it makes sense to let the assignment happen on a much faster time scale. Also delays before assignments, if taken into account, do not change the balancing behavior of the approach. The second assumption is a bit stronger, as it does not consider the effect of the delay between vehicle assignment and arrival at the charging station; however it may be a valid approximation if the partition cell is small and distances to the charging station can be covered much faster than it takes to charge a vehicle. The third assumption is purely to simplify notation. It is possible to compute the exact expression, which will include higher order terms and exchange the simpler expression for it. Assumptions (iv) and (v) are quite standard in queueing theory, see for example [165] for a similar model.

Under the above assumptions, the system can be modeled as a Markov chain, where for n_m charging stations the state of the Markov chain is an n_m dimensional vector that reports the occupancy of each station. The transition

probabilities between the states can be readily computed from the rates of
the arrival process and the service process and the used assignment rule for
each station. For the algorithms described above it immediately follows that
the Markov chain is irreducible and primitive [141], which assures existence
and uniqueness of a stationary distribution. If the number of customers in
each station is bounded, then the number of states is finite and we can use
the transition matrix to compute the stationary distribution, which yields
information about what fraction of time the chain spends in each state.

6.4.2 Quality of Service Analysis: Waiting Times

We now give some complementary results to indicate how quickly EVs are
assigned to a place in a queue at a charging station. We call this time the
waiting time. As before, to facilitate exposition, we make the following assumptions. (i) If two EVs sense a green light, then only one EV is assigned to
the available charging station, for instance the one with the minimum residual
charge. Similarly, if one EV senses multiple green lights, then it is assigned
to only one charging station (e.g., the closest one). (ii) We assume that the
probability that an EV will be assigned to charging station i is

$$\frac{p_{CP}^{(i)}}{\sum_j p_{CP}^{(j)}} \tag{6.3}$$

which is obtained by neglecting high order terms (i.e., considering only terms
that are linear with respect to the probabilities). (iii) In deriving the following
lemmas we make a stationarity assumption; namely, that the system has
reached steady state and the number of vehicles queuing at each of the
charging stations is constant.

Lemma 6.4. (EV waiting time) : *Consider a network of n_m identical
charging stations, available for charging at all times. Assume that at a certain time the ith vehicle makes a request to access a queue for charging at a
whatever charging station. Then the average waiting time $t_{wait}^{EV_i}$ before being
assigned to a specific charging station is*

$$t_{wait}^{EV_i} = \left(p_{EV}^{(i)}\left(1 - p_{red}\right)\right)^{-1} \tag{6.4}$$

where

$$p_{red} = \prod_{j=1}^{n_m} \left(1 - p_{CP}^{(j)}\right), \tag{6.5}$$

$p_{CP}^{(j)}$ *is the probability that the jth charging station broadcasts a green light, and
$p_{EV}^{(i)}$ is the probability that the ith EV senses a green signal, as in Algorithms
6.1 and 6.2 respectively.*

Proof. At any one time step the ith EV gets assigned if it receives at least one green signal and it is currently listening for signals. These events occur with probability $1 - p_{red}$ and $p_{EV}^{(i)}$ respectively. As the two events are independent, the joint probability is $p_{EV}^{(i)}(1 - p_{red})$. From this it is possible to compute the expected waiting time, which can be seen to be equal to the one given in Equation (6.4). □

According to Lemma 6.4 the expected waiting time is short if p_{red} is small, which occurs if the number of charging stations is large or if the queues are small. Lemma 6.4 can be interpreted as a quality of service measure for the customers. One may equally be concerned about the quality of service for the charging stations. For the charging stations it is undesirable to be idle. Obviously this situation will occur at times with low demand. However, the system should avoid having one or several stations idle, while other stations have a high workload. The following lemma quantifies the probability that the assignment rule assigns cars in a non-optimal way.

Lemma 6.5. *Consider a network of n_m identical charging stations. At a given time, let $\mathcal{B}$ be the set of all stations that are busy charging one or more vehicles, and let n_I be the number of charging stations that are idle. Assume that at this time instant a vehicle is assigned to one of the stations. Then the probability that this vehicle is assigned to a busy station instead of an idle one is given by*

$$1 - \frac{n_I}{n_I + \sum_{i \in \mathcal{B}} F(N_i)}. \tag{6.6}$$

Proof. According to our assumptions the new car is assigned to one of the busy stations with probability

$$\frac{\sum_{i \in \mathcal{B}} F(N_i)}{\sum_{j=1}^{n_m} F(N_j)}. \tag{6.7}$$

Further $F(N_j) = 1$ for any idle station. Using this in Equation (6.7) and rearranging it yields Equation (6.6). □

Remark 6.6. Lemma 6.5 addresses the undesirable situation where a car is assigned to an occupied station even though an idle station is available. It is possible to obtain more comprehensive quality of service measures along the same lines as in the lemma. For example it is possible to compute the probability that a newly arriving car is assigned to a specific charging station. Such problems can be addressed as a queueing theory problem in an almost identical manner to Lemma 6.5. Given the exact shape of the function F it is possible to further refine this result. For example, if one uses a function of the form of (6.2) the probability of a new car being assigned to a particular charging station depends not on the absolute values of the queues, but only on the difference of the queues' lengths.

In Lemma 6.5 the probability that a new EV is assigned to a busy station instead of an idle station depends strongly on the shape of the function F. Here a rapid decrease of F is preferable. For example, consider a system with 10 charging points, of which 2 are idle. Assume that all other stations each have 2 EVs in their queue. If we use (6.2) to define F, then the probability given by (6.6) is less than 0.04.

6.5   Simulations

We now present two simulations to illustrate the efficacy of the approach. For our simulations we used again Equation (6.2) to calculate $p_{CP}^{(i)}$.

Simulation 1 (Deterministic vs. Stochastic) : The first simulation compares the stochastic algorithm with a deterministic one. The simulation runs for 8 hours. Moreover, we assume a fixed charge rate of $0.01\,\mathrm{kW/s}$ per vehicle and a charge request of energy ranging between 1 and $10\,\mathrm{kWh}$, so that the maximum time for recharge (i.e., corresponding to a request of $10\,\mathrm{kWh}$ of energy) is 1000 seconds (i.e., ≈ 16 minutes). We also assume that at the beginning of the simulation (8 AM) all the 20 available charging stations are empty. In the deterministic approach a new car is directly assigned to the charging station having the shortest queue. Figure 6.1 and Figure 6.2 compare the results of both approaches.

Remark 6.7. It can be seen from Figure 6.1 and Figure 6.2 that the deterministic assignment strategy is better than the stochastic approach. By its very nature, the deterministic method is optimal and achieves perfect balance more often than any other assignment algorithm. However, the aim of our work here is to validate the performance of the stochastic approach, which may be desirable for a number of reasons. For example, the deterministic method requires a large amount of communication between all participants (cars, stations, cloud). With this in mind Figure 6.1 shows that the average number of vehicles waiting at the charging stations using the stochastic algorithm is very close to that obtained from a centralized solution. The two approaches also provide very close results in terms of the variance of the vector of queue lengths (variance equal to zero corresponds to exactly the same number of vehicles queuing at each charging station), as shown in Figure 6.2.

Simulation 2 (Plug-and-play Behavior): A feature of the proposed approach is that it provides the opportunity of handling a dynamically varying number of available charging stations in a manner that is completely transparent to both the vehicles and the stations. Such a possibility is convenient in several circumstances, for instance shopping malls or workplaces might reserve charging facilities to customers and employees during working hours and

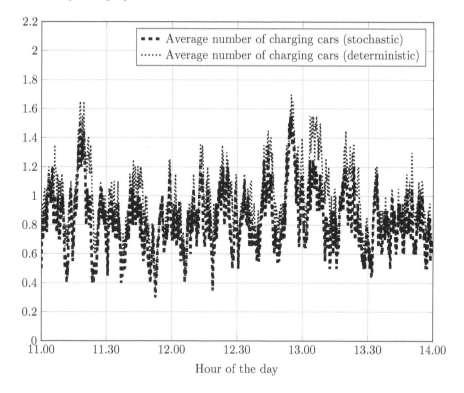

FIGURE 6.1
The decentralized minimum-communication stochastic approach provides results very close to the heavy-communication deterministic centralized approach: here the average number of queuing vehicles at each charging station is shown.

be publicly available at other times. In this simulation we show that the algorithm can be used to achieve such functionality. We assume that we have three available charging stations until 12 PM, at which point two new charging stations start offering charging facilities.

Figure 6.3 illustrates that the maximum difference between the longest queue and the shortest one is usually very small (well-balancing), and is particularly large only when the new charging stations become available. As depicted in Figure 6.4, as the new two charging stations become available, very short waiting times are restored.

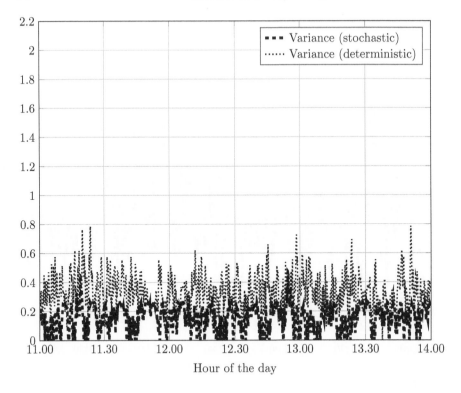

FIGURE 6.2
The decentralized minimum-communication stochastic approach provides results very close to the heavy-communication deterministic centralized approach: here the average variance of queuing vehicles at each charging station
is shown.

6.6   Concluding Remarks

This chapter illustrates an approach to reduce the potential for queuing of
electric vehicles at charging stations based on an analogy between electric
vehicles/charging stations and mobile phones/base stations. The second contribution is a stochastic technique for balancing load across queues. The resulting system balances load across the charging network, and avoids peaks
in individual queues, while at the same time avoiding driver inconvenience
through enforced route selection.

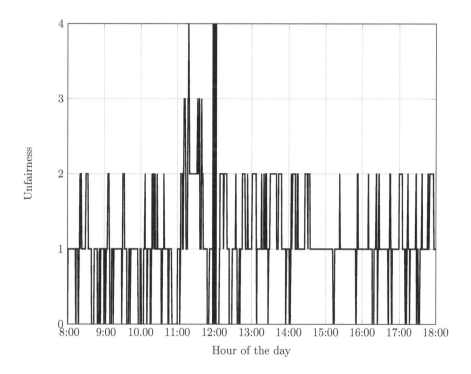

FIGURE 6.3
The fairness of the algorithm is measured in terms of maximum difference
between the busiest and the most free charging station (larger than 1 implying
a sort of unfairness).

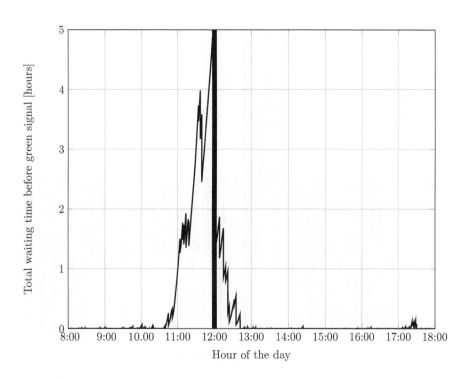

FIGURE 6.4

Aggregate time required before receiving a green signal. As can be seen, the waiting time is greatly reduced when the 2 new charging stations are available, which occurs at the time indicated by the vertical line.

7

Charging EVs

7.1 Introduction

Power demand usually follows periodic patterns, with a higher demand of power during the day, and a lower demand during the night, when many activities are stopped. The exact shape of the electrical load pattern depends then on a number of factors, such as the location, the day of the week, the season or the weather. At all times, power demand has to be matched by an equally high power generation. In a classical distribution grid, there are several grid services responsible to achieve this match on various time scales. When the power demand is very high, i.e., at peak times, the power grid operates closer to its physical limits, and this implies that usually energy generation is more expensive during these periods. Another consequence of these peaks is that the transmission losses are higher, since the losses are proportional to the square of the current, while the demand is proportional to the current. Hence, to improve network utilization, it would be desirable to reduce peaks, possibly by shifting power consumption to moments when the demand is low, i.e., valleys, and eventually reshaping the demand curve to make it as flat as possible [29, 68, 1, 63, 118]. Consequently, the charging of EVs can increase the generation costs dramatically when the charging occurs during peaks, or even cause power quality issues, such as voltage deviations, see for example [154, 51, 68, 146, 155]. In this regard the power grid could even become a limiting factor in the adoption of EVs [146, 155, 151]. In addition, the distribution of the EV charging, in a worst case scenario, can also lead to imbalances in the three phases of the network loading, and negatively affect the voltage profile and in general power quality [150].

While the previous considerations may be seen as an impediment to a smooth adoption of EVs, there is also a general expectation that a smart utilization of EVs may actually eventually support the operation of the grid. In particular, this is due to the fact that, nowadays, the power grid is undergoing a deep transformation to facilitate the inclusion of Distributed Energy Resources (DERs), especially from renewable sources, i.e., wind and sun. DERs are intrinsically small-sized, non-dispatchable and intermittent power plants, and their widespread adoption is changing the classical top-down structure of power grids. In particular, tighter stabilizing measures are required to increase power generation from DERs ([64, 160]), and other expensive infrastructure, i.e., batteries or super capacitors, is used to increase the flexibility of DERs. In addition, ancillary services such as load management or demand side management, are becoming popular as an alternative to the classical paradigm that power generation should follow the demand, and rather, the demand is controlled in order to follow the generation. Accordingly, some electric loads adapt their instantaneous power consumption to fulfill some ancillary services that can be useful for the grid [108]. For example, excess wind energy generated at night is stored in form of thermal energy in refrigerator units such that later during peak times this energy can be used to reduce the output of a conventional coal power plant [191]. In this context, while not all electrical loads are feasible candidates for such demand side management services, EVs are ideal candidates, since they fulfill three key requirements, as identified by [25, 29, 64].

- **Flexibility:** The charging of an EV can be scheduled with some degrees of freedom, as it can be often postponed and/or reduced for a significant period with small inconvenience to the owners.

- **Predictability:** Once connected, it is possible to accurately predict the energy needs of an EV. Further, it is possible to estimate in advance the long term energy requirements using historical driving patterns [151].

- **Availability:** If EVs are connected to the grid whenever they are idle, including at home and at work, a considerable number of EVs would be available throughout most of the day, thus forming a virtual battery of considerable size.

The possibility of using EVs as storage devices implies that in principle, if needed, EVs can also inject power back into the power grid. Such an operation is commonly known as V2G [1, 147], as the power flows in this case from the vehicle to the grid. Although V2G operations have the potential to support the power grid tremendously, they should be performed in a cautious manner. In fact, V2G services may have some implications for customers. Firstly, the lifespan of the battery of the vehicle may be reduced due to the additional stresses caused by the charging cycles of V2G services, that may possibly also include over-depletion cycles. Secondly, if the owner collects the vehicle earlier than originally planned, V2G services may not leave enough energy for the next trip. Finally, one should also pay attention to environmental consequences, as we will further discuss in Chapter 8.

There is also a third enabled service, that is usually less discussed but also attractive, that is enabling the EVs to participate in reactive power balancing tasks. This service consists in supporting the power grid by delivering or absorbing reactive power to improve the power factor in a region. It is expected to be more useful for vehicles parked close to industry sites, where more reactive power is produced or required. The advantage of using EVs to counteract reactive power demand, is that it will not be required to transport reactive power over long distances, which in turn reduces not only the costs and transmission losses, but also increases the amount of active power that can be transmitted to the loads.

From the previous discussion, it is clear that EVs might actually support the grid and even facilitate the increased deployment of renewable energy generation. However, there are some obvious limitations that need to be considered. These are:

- **Limited energy**: Each EV can only draw, or provide in case of V2G operations, a limited amount of energy due to its limited battery capacity. Naturally, the trend to increase the battery sizes and the increase of penetration levels will mitigate this limitation.

- **Limited power**: The charger, as well as the battery, of an EV will have a limited power that can be handled. Much of the controlled EV charging will most likely occur at home where a single phase outlet is used, with a capacity around 4 kW [37, 51, 68]. In the future this might change and more three-phase charging with higher power capabilities will be used. As for the limited energy availability this development and increased penetration levels mitigate the effects.

- **Limited connectivity to the power grid**: Although it has been estimated that cars are on average parked up to 90% of the time ([31]), not all parking spots are equipped with plugs that would actually connect the vehicle to the grid for Grid to Vehicle (G2V) or V2G power exchange. In this context, an increase of charging capabilities at work places, shopping centers and on street parking, would significantly increase the availability.

- **Customer constraints**: It is important to consider the constraints of the owners in regard to their transportation needs (e.g., in terms of the desired driving range or utilization of additional energy-demanding on-board services). These needs can vary between persons, charging locations, and charging times.

- **Limitations of the charging cycle**: Very often a Constant Current, Constant Voltage (CCCV) method is used to charge EV batteries, where first a constant charging current is applied until a certain voltage is reached. Afterward, the voltage is kept constant while the charging current is decreased. While during the constant current mode it is possible to adjust the current (and thus, control the power), this cannot be done during the constant voltage part. Hence, the EVs should be treated as uncontrollable loads during part of the charging process. Also, especially when the voltage of the battery is low it is recommended to use low charging currents [119].

7.2   EV Charging Schemes

Due to the previously described implications, controlled EV charging is an active research area. There is a vast literature concerned with EV charging, see for examples [51, 1, 147, 36, 68, 196, 143]. The suggested methods vary widely throughout the literature. For comparison purposes of these methods, the following properties can be identified: (i) control architectures; (ii) communication requirements; (iii) degree of control actuation; (iv) supported services; (v) control methods; (vi) measurement and prediction requirements; (vii) operational time scales; and (viii) control focus. In the following, we shall give a brief discussion of the existing literature organized in terms of these characteristics.

7.2.1   Control Architectures

Commonly three control architectures are employed for charging plug-in EVs.

- **Centralized control:** A central controller aggregates information from connected EVs, and possibly from connected measurement devices, and then uses these measurements to make a decision in regards to the charging. The control decision is then communicated to the connected EVs, which will make the necessary low-level adaptations to their charging rate.

- **Distributed control:** In this case there exists a central management unit that can gather information from measurement devices and/or connected vehicles. This information is then communicated to the connected vehicles that use it to augment local information and take appropriate charging decisions.

- **Decentralized control:** In this case, coordination is achieved using purely local measurements and communication among participants. Each participant, as in the distributed case, chooses the appropriate action given the available information.

Many studies assume a centralized control structure, see for example [132, 51, 37, 147, 177]. However, using a central control structure implies that the central management unit requires large amounts of information to take the appropriate charging decisions. In the case of EV charging this can include a large amount of data such as the energy needs of every single vehicle, estimated connection and disconnection times, as well as the physical limitations imposed by the charger and the battery. Note that such an information exchange may in some cases not be desired, or even possible, due to communication limitations or privacy issues. After gathering all the information, the centralized controller will take the appropriate decision as to which control action that should be taken, i.e., how to schedule the EVs. This decision can be taken using various methods. For example, in the centralized control structure of [51], vehicle owners may select a priority for their preferred charging time during the evening and night, where the highest priority coincides with the evening peak. Then, the algorithm schedules the charging times of the vehicles according to the selected priorities and the impact on the distribution grid, guaranteeing that the distribution grid will operate within its allowed constraints. Power flow analysis is continuously performed to keep checking that the limits imposed by the distribution grid are not violated. Alternatively, another frequent solution is to define an objective function and optimize the scheduling to minimize (or maximize) the objective function. Such an approach is taken by [177, 132, 89]. For example, in [132] the objective function depends on the SOC, the battery capacity, the expected charging time, as well as the willingness to pay more (or less) by the vehicle owner. After the optimization is concluded, the centralized controller informs the vehicles about their schedule.

The main disadvantage of centralized control structures is that they are generally less robust to failures of components and often scale badly with the size of the problem. For example, the strategy used in [51] requires repeated load flow analysis for each EV connected to the distribution grid. Similarly, strategies based on centralized optimization may in some cases become prohibitively complex to solve.

Despite these disadvantages, central controllers can usually find a globally optimal scheduling, since all the required information is available to the single centralized controller. However, this raises other issues in regard to data protection. The second main advantage of such controllers is that it is possible for the provider to adapt the behavior of the control easily from a central location. For example, a change in the desired demand can be easily achieved by properly adjusting the central controller.

An alternative architecture is the decentralized control structure. Its main advantage is that it is usually more robust to component failures, and little dedicated infrastructure is required. However, the achieved performance is often sub-optimal, because not all the information is available to every single car in the network. Additionally, adjusting the behaviour externally can prove more challenging than in a centralized approach, due to the individual controllers of each EV.

Distributed control structures combine the robustness of decentralized algorithms with the ability to easily govern the overall behavior from a central management unit, as in centralized control structures. For this reason, most algorithms designed for the control of EVs are distributed algorithms, see for example [196, 68, 60, 59, 118, 172, 171, 176, 173, 170, 74].

7.2.2   Communication Requirements

A first concern regarding the communication aspects, is that exchanging large amounts of data may lead to considerable communication and computation delays, especially for large scale systems like power networks. Such delays can degrade the performance, and eventually cause damage to the power grid; for example, if transformers are overloaded for too long periods. This is especially critical for fast reacting services. Further, while it is generally acknowledged that distributed and decentralized algorithms impose lower communication loads than centralized ones, this is not always the case. For instance, distributed algorithms with a high communication load can be found in [132, 196], while efficient ones can be found in [68, 60, 59, 172, 171, 176, 170].

A second concern involves privacy issues and data protection. The more data that is shared, the more important it becomes to prevent misuse of the information. In this context, it is not necessarily important how much data is transmitted, but also what data is exchanged. This varies widely among the proposed schemes. For example, the centralized algorithm suggested in [51], needs a two-way communication between the central controller and the EV. The vehicle transmits a preferred charging time and a load profile of the vehicle, as well as its connection and disconnection times. In return, the central controller transmits the required charging allowance. Even though the communication load is relatively low, this algorithm does not scale efficiently with the number of EVs, and the transmitted information may be intercepted by malicious agents. It might be also possible to use no communication at all among the agents and utilize the frequency and voltage as an indication of the required actions. While to the best of our knowledge we are not aware of an EV charging mechanism doing this, there is an example of such an algorithm for thermostatically controlled loads in [11].

7.2.3   Degree of Control Actuation

The next important characteristic is the ability of a controller to manipulate the charge rates. Very roughly, it is possible to distinguish three main categories: binary charging capabilities (i.e., on-off charging), multi-step charging, and continuously adaptable charging capabilities. Generally speaking, a higher flexibility in actuation leads to a better control over the aggregated power consumption [132]. However, additional flexibility increases the complexity as well as the cost of the infrastructure, such that the increased performance has to be weighted against these costs. Many recent works propose algorithms that require binary charging capabilities, such as [147, 51, 176]. Note that in [1] the EVs at electrical nodes are aggregated and that this aggregated power can vary continuously within given limits. Also, later in Section 7.3 we shall describe a specific framework that can exploit both binary and continuously adaptable power levels.

7.2.4 Supported Services

A plug-in EV can, in principle, support three basic charging services: uni-directional power flow adjustments; bi-directional power flow regulation, i.e. V2G operation; as well as reactive power exchange with the grid. The first mechanism is the basic property needed for controlled charging, see for example [36, 143, 196, 68, 37, 51, 172, 171, 176, 170, 132, 154, 59]. This basic capability is considered necessary for a widespread deployment of EVs [146, 155, 10, 143, 150, 73, 37, 63]. Other studies have also investigated the feasibility of V2G services, see [89, 1, 147, 173, 170] for examples of this. An example of how a charging algorithm can be extended to include V2G services is given in Chapter 8.

As mentioned, such services can bring additional problems for the EVs in the form of additional stresses on the battery and eventual implications on the customer or the environment. For a discussion on the latter topic we will refer the reader here to Chapter 8. To lessen the impact on the vehicle owner [1] imposes limitations on the energy that can be used for V2G operations and/or the time when such operation is allowed.

The third service that can be provided by EVs is reactive power exchange. While this capability is not widely considered in the literature, it seems to be a logical next step beyond conventional charging. Most likely, it will be adopted and useful for areas with high penetration level of EVs that are in close proximity to large consumers and providers of reactive power. For instance, work place charging of vehicle fleets might be a candidate example for such a service [31, 173, 170].

7.2.5 Control Methods

In the load management literature there are three main ways to control power consumption [188, 112]:

- **Time of use:** Here, power consumption is indirectly controlled by varying the electricity price, depending on the time of the day. Prices are usually fixed and there are two or three different levels of price. Such schemes are already in use in many countries. For example the electricity provider AGL in Australia defines three such levels: Peak, Shoulder, and Off-Peak[1].

- **Real time pricing:** Here, the electricity price varies in real time depending on the demand and the generation. It is expected that the customers, or the electric appliances themselves, will react to such a varying price with an equally fast adaptation of the charge rates. The response to these

---

[1]AGL Energy Price Fact Sheet, available online at http://www.agl.com.au/~/media/AGLData/DistributorData/PDFs/PriceFactSheet_AGD20308MR.pdf. Last Accessed July 2017.

prices can be automated to achieve an optimal behavior.

- **Direct load control:** Here, the charge rates are directly adjusted by the load management scheme without the need for pricing adjustments or pricing signals. Naturally, participation might be encouraged by monetary incentives.

The authors of [112] discuss various other methods of control, such as load curtailment programs, other pricing programs, or education programs. However, the above three types are the most popular in the literature and we will therefore focus on those.

Control methods based on the time of use are already exploited by some electricity providers. While they are effective in shifting part of the load away from peak times (due to the higher prices), their effectiveness is limited. The main reason is that these are indirect methods that require individual actions from the owners. Further, they are relatively inflexible due to the fixed time periods. In addition, even if a charging action is shifted to off-peak times, there may be a second peak later when all EVs commence charging at the start of the off-peak pricing period [151]. Hence, most works actually recommend methods either based on real time pricing [68, 196, 60, 118, 59] or on direct control [37, 51, 1, 29, 132, 154, 103, 172, 171, 176, 170]. On the other hand, real time pricing causes the EVs to react automatically by optimizing a specific cost defined by the algorithm. Firstly, this requires the customers to trust that the actions undertaken are in their best interest. Secondly, it shifts the burden partly towards the customers, since they are charged higher prices for the consumed power during peak demand. Finally, also direct load control is considered controversial since the customer's needs might not be taken into account. For instance, if the grid remotely disconnects some EVs to reduce the peak, then a vehicle owner may not have enough energy for his or her next trip. One way to handle this drawback for the owners is to introduce constraints. For example, in [51], the vehicle owner selects a preferred charging time, while in [37] the vehicle owner dictates a charging deadline when the vehicle has to be fully charged. In other situations it might be not essential to fully charge a vehicle, since the charging will be an additional service provided to the customer, see for example [172, 170].

7.2.6   Measurement and Forecasting Requirements

Most charging schemes exploit some additional forecast information, such as daily demand predictions [68, 37, 132] or energy requirements of the EVs [68]. Robustness to errors in such predictions is an important factor, since predictions cannot be expected to be perfect.

7.2.7 Operational Time Scales

The previously discussed load management tasks can operate at various time scales. The specific time scale will depend on the restriction caused by the communication network as well as the capabilities of the chargers themselves. Hence, the proposed smart charging algorithms vary in the implemented time scale. While 15 min time slots are considered in [37, 154], [51] uses 5 min slots, [68] assumes in their simulations hourly time slots, and [12] argues for the possibility of very fast time scales of a few milliseconds. Other studies allow for a range of different time scales [172, 171, 176, 170, 68] without specifying the actual length.

7.2.8 Charging Policies

Many studies compute the power allocated to EVs to mitigate the inconvenience to the grid as much as possible, given the charging constraints (or preferences) of the EVs. Alternatively, the EVs can adapt their charging to maximize customer satisfaction, or in other terms the Quality of Service (QoS) to the vehicle owners, given the constraints of the power grid, see for example [51, 170]. Both optimization perspectives have been investigated in the literature, and lead often to two different outcomes. In the first one the actual charging demand depends on the individual choices, and may still result in adverse effects on the grid. On the other hand, the second perspective may give rise to some inconveniences to the customers, since the hard constraints are dictated by the distribution grid.

7.3 Specific Charging Algorithms for Plug-In EVs

In the following, let C_p denote the available power in the resource sharing problem. This available power is specified by the energy provider as the desired maximum demand by electric loads. While in some cases these electric loads will consist solely of EVs which charge in a controlled manner, it might in other cases include other non-controlled EVs or other electric loads. We will assume that the charging algorithms operate with a slotted time frame, so that changes in operation occur at discrete time steps. The time steps are indexed by integers k. Further, we let $P_i(k)$ be the power drawn by EV i at time step k and $n(k)$ be the total number of vehicles participating in the charging scheme. Note that the number of participating EVs is dependent on k since vehicles can connect and disconnect from the distribution grid. In what follows we assume that there are no non-controllable loads. Then, the power constraint can be expressed by

$$\sum_{i=1}^{n(k)} P_i(k) \leq C_p. \tag{7.1}$$

In these charging tasks a second constraint imposed by the individual chargers comes into play. This means that the power drawn by an EV cannot exceed the rated power output of the charger supplying the vehicle $\overline{P}_i(k)$. For instance, in some domestic scenarios, that is where plug-in EVs charge at a single phase home outlet, we would choose $\overline{P}_i(k) = 4\,\text{kW}$, since the maximum power output of a standard single phase 230 V outlet is 4.6 kW [37]. In addition, the controller will typically limit the charging to lie within a predefined feasible set $\mathcal{P}$, i.e.

$$P_i(k) \in \mathcal{P}. \tag{7.2}$$

For example, for chargers that can only switch the charging on or off, this means that $\mathcal{P} = \{0, \overline{P}_i\}$. Note that we here neglect any power consumed by the charger. Given these constraints the EVs try to maximize the power taken from the grid (i.e., Equation (7.1)) subject to the user-centric constraints (i.e., Equation (7.2)).

7.3.1   Management Strategies

The system underlying the proposed charging structure consists of three parts: the power grid, the EV chargers, and a central management unit. These three parts are interlinked through the exchange of power and/or information. The power grid connects the feeders, power generators, and the charging vehicles. At the same time the central management unit has access to power grid state information, such as the power flow at a transformer or power line. It uses this information to decide when to send broadcast signals to the connected EVs. These EVs then react individually. Note that the central management unit does not receive any individual information from the EVs. Neither, do the EVs communicate among themselves. Hence, there is very little critical information exchange, which lessens concerns regarding data protection and privacy issues. However, a key requirement is that the infrastructure is able to communicate to vehicles. While such functionality is not widespread, the assumption of grid-to-vehicle communication underpins most of the work in this area. The central management unit monitors the current demand and is informed about the available power from the grid. As soon as the demand exceeds the available power, i.e., constraint (7.1) is violated, it sends a broadcast signal to inform the connected vehicles that the power consumption should be reduced. We will call such a signal, *a capacity event* signal. The reaction of the EVs to capacity event signals is determined by an algorithm that runs independently locally on each connected EV. This algorithm is the key part of the management system and defines the behavior of the overall distributed charging system. In the remainder of this chapter we shall describe two such algorithms.

7.3.2 Binary Automaton Algorithm

The binary automaton algorithm controls the demand of the vehicles by interrupting their charging progress throughout the time they are connected to the charger. This means that at certain times while the EV is connected, it interrupts its charging progress, i.e. there is no power demand, until it is allowed to continue. At regular intervals, indexed by k, the central management unit will send capacity event signals or refrain from sending a signal, depending on current overall consumption. The EVs react to these signals in a different manner depending on whether they are currently charging or waiting to be allowed to start charging (again). Note that we here assume that once the EV finishes charging, i.e. its battery is fully charged, it disconnects from the charger.

The "turn-off" phase is executed if a capacity event signal has been received. In this case some of the EVs that are currently charging will decide to interrupt their charging progress. Additionally, during this phase no EV is allowed to start or resume charging, if they are currently waiting. Hence, during this phase EVs either continue uninterrupted with their charging or turn off, i.e. stop consuming power. In this way the aggregated demand by EVs is likely to be reduced as desired. The decision whether or not to interrupt the charging is decided by the EV in a probabilistic manner and is governed by two variables: an individual probability $\mu_i(k)$ and a constant integer s. The integer s is a common variable among all agents and hence has to be somehow communicated among the EVs. Since it is constant this can however be directly fixed in the charger upon the installation. Then, with probability $\mu_i(k)$ the EV will interrupt its charging, its power consumption is set to $P_i(k+1) = 0$, and it resets the probability $\mu_i(k+1) = \frac{1}{s}$. On the other hand, with probability $1 - \mu_i(k)$, the EV decides not to interrupt its charging, its power consumption is set to $P_i(k+1) = P_i(k)$, and the EV increases the probability of interrupting for the next time step by setting $\mu_i(k+1) = \mu_i(k) + \frac{1}{s}$. This also means that after at most s time steps with a capacity event the EV will interrupt its charging progress.

The "turn-on" phase is structured in a similar fashion. It governs starting or resuming, if previously interrupted, charging and is executed whenever no capacity event signal is received. A vehicle that is already switched on is not affected by this phase and continues uninterrupted. As before each EV maintains two variables that are associated with this phase: the probability $\nu_i(k) \in [0,1]$ and an integer r common among all EVs. With probability $\nu_i(k)$ the vehicle sets its power consumption to $P_i(k+1) = \bar{P}_i$ and resets the probability to its initial value, i.e. $\nu_i(k+1) = \frac{1}{r}$. So in this case the EV starts charging or resumes doing so, if it was previously interrupted in a turn-off phase. With probability $1 - \nu_i(k)$ the EV remains switched off. In this case it increases the probability of turning on by setting $\nu_i(k+1) = \nu_i(k) + \frac{1}{r}$.

Figure 7.1 depicts a detailed flow chart of the algorithm as executed at each time step by the EVs. Note that the common parameters r and s define

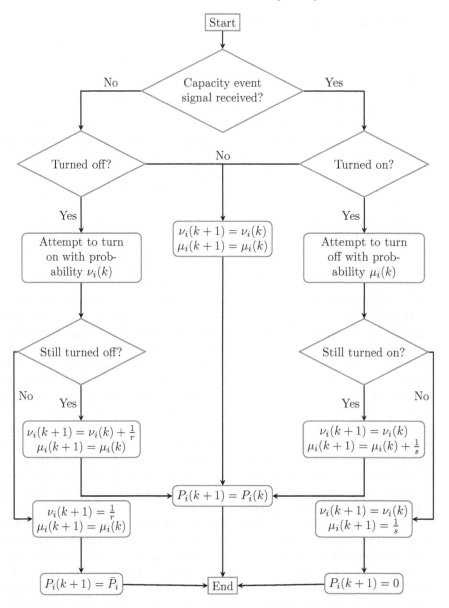

FIGURE 7.1
Flow diagram of the binary automaton algorithm executed by a vehicle to
control the charging procedure

the behavior of the algorithm. The larger these are chosen the longer the EV
is expected to remain on or off. In this simple setting we assume that a fair

share is achieved if the duty cycles of the vehicles are identical on average, i.e., the time they are on is on average the same for all EVs. This is achieved by selecting r and s identical for all EVs.

For a detailed mathematical analysis of the algorithm, we refer the reader to [176, 169].

7.3.3 AIMD Type Algorithm

Additive Increase Multiplicative Decrease (AIMD) algorithms, see [34] and Chapter 18, have been very successful in regulating congestion in the Internet [167]. Bandwidth sharing in the Internet presents similar issues to power sharing at a charging station: flows leave and join the network; the available capacity may change over time; and the system can be very large scale. AIMD algorithms are characterized by an Additive Increase (AI) phase and by a Multiplicative Decrease (MD) phase. Accordingly, during the AI phase the EVs increase their charge rates linearly with time, by an additive constant $\alpha > 0$ (in $\mathrm{kW\,s^{-1}}$), as shown in Figure 7.2. When a capacity event signal is received, each EV reduces its drawn power by a multiplicative factor $\beta \in (0,1)$. Note that the capacity event in our case corresponds to the congestion event in the case of Internet flow control. While there the congestion event point to a congested link, here the capacity event indicates that all available power is used. A more detailed discussion of the AIMD algorithm and its properties is given in Chapter 18. A slightly more complex variant of the multiplicative

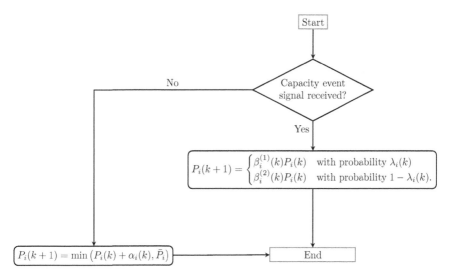

FIGURE 7.2
Block diagram of the AIMD algorithm. The algorithm is run by each EV independently.

decrease is the probabilistic selection of the reduction factor. This variant
helps maximize the transferred power, while still satisfying the constraints on
total and local powers. In our case, upon detecting a capacity event, we assume that each vehicle can either reduce its charge rate by a large or a small
amount, according to two possible multiplicative factors denoted by $\beta^{(1)}$ and
$\beta^{(2)}$. The choice of β is determined in a probabilistic manner where we use λ_i
to denote the probability that EV i chooses $\beta^{(1)}$ (and thus, $1 - \lambda_i$ for $\beta^{(2)}$).
Controlling these probabilities as well as the other parameters α, $\beta^{(1)}$, and
$\beta^{(2)}$ enables us to implement different policies among the EVs. For example,
in some situations policies to minimize cost are important, and in other situations speed of charging is important. To give a flavor of how different policies
might be realized we consider two typical scenarios: a domestic charging scenario and a workplace scenario. Note that other scenarios can be found in
[172, 171, 173, 170]. In each situation the fundamental assumption is that the
available charging power must be shared by a number of vehicles to achieve
certain policy objectives.

7.4   Test Scenarios

We use the following use cases to test both algorithms.

7.4.1   Domestic Charging

In a domestic scenario the available power is shared among the connected
EVs without taking the vehicle charging time into account. This is typical of
night-time charging when there are no real time limitations, and when each
of the users subscribes to a flat-rate contract. So the objective is to maximize
the transferred energy to the vehicles, while assigning the same priority to
each vehicle. This is in its form similar to the fairness notion in the binary
case, however we aim to equalize the charging rates. This can be achieved by
selecting the parameters α, $\beta^{(1)}$, $\beta^{(2)}$, and λ identical for all EVs.

7.4.2   Workplace Scenario

One property of a workplace scenario is that EVs do not *compete* but *cooperate*
to achieve a common goal. This implies that an EV is willing to receive a
smaller charge rate than another EV in the case that the second EV has more
need for energy. We assume here that upon connection, each EV will request
the desired quantity of energy and will also specify the expected deadline for
recharging. Then, we denote by $\underline{P}_i$ the minimum power demand to serve an
EV as requested without exceeding the power limitation of the charger $\bar{P}_i$.
For instance, if EV i requires $16\,\mathrm{kW\,h}$ of energy in $8\,\mathrm{h}$, then $\underline{P}_i = 2\,\mathrm{kW}$ if

the maximum charge rate is above 2 kW. The infrastructure will then try to give as much power as possible to each EV, keeping the shares of power proportional to the requested charge rates $\underline{P}_j$. To achieve this, we tune the additive parameters α_i to be linearly proportional to the desired charge rates $\underline{P}_j$. In particular, to minimize EV communication, we assume that an off-line map can be used by each vehicle to compute the α_i given its own $\underline{P}_j$. Here, we assume that the maximum $\alpha_i = 0.1\,\mathrm{kW\,h^{-1}}$ corresponds to the maximum $\bar{P}_i$, while $\alpha_i = 0$ corresponds to a zero charge rate. Intermediate values of α_i can then be found accordingly. AIMD does not require the knowledge of the total available power, and thus the algorithm can be implemented by each EV *without any communication at all*, except that of a broadcast of the capacity event notification.

Remark 7.1. The use of proportional α_i to achieve linearly proportional P_i is motivated by the fact that the fixed point P_i^* of the dynamics for AIMD with linear constraints (but no constraints in the single P_i), if any, is given by

$$P_i^* = \Delta\tau(1 - \beta_i)^{-1}\alpha_i, \tag{7.3}$$

where $\Delta\tau$ is the (possibly average) time between two successive capacity events.

7.5 Simulations

We will give here simulation results using MATLAB® for various scenarios in two simulation settings. For the sake of illustration, we neglect here the distribution grid and any non-controllable loads. For interested readers we refer to [169, 170], where these algorithms are simulated in a power grid using the standard IEEE37 test feeder[2].

The first simulation setting assumes a constant available power C_p of 40 kW and a fixed number of 25 EVs connected. They are all connected from the beginning of the simulation, with a random request for energy, until they receive the required energy. The second simulation setting assumes a dynamic environment where EVs connect to the grid at any time, with a random request for energy, and remain connected until fully served. The available power C_p varies randomly with a rate limitation of $5\,\mathrm{kW\,s^{-1}}$. The total number of EVs that are allowed to connect is 25. However, as mentioned before they connect randomly. The connection times are drawn from a uniform distribution, however we limit the connection time to the first hour of the simulation.

Many parameters are identical for both settings. The simulated period is 4 h with a time step duration of 1 s. Each EV has an energy requirement

---

[2]Distribution Test Feeders, available online at https://ewh.ieee.org/soc/pes/dsacom/testfeeders/. Last Accessed July 2017.

between 2 kW h and 10 kW h, which is drawn from a uniform distribution. The maximum power consumption of an EV is set to 3.7 kW.

In total we investigate four scenarios:

1. Binary Algorithm: All connected EVs utilize the Binary Algorithm to achieve a fair share of their on times.

2. AIMD in a domestic scenario: All connected EVs utilize the AIMD algorithm where their parameters are selected as in Section 7.4.1.

3. AIMD in a workplace scenario: All connected EVs utilize the AIMD algorithm where their parameters are selected as in Section 7.4.2.

4. Binary Algorithm and AIMD in domestic scenario: EVs utilize either the Binary Algorithm in Section 7.3.2 or the AIMD algorithm where their parameters are selected as in Section 7.4.1.

7.5.1   Binary Algorithm

As mentioned the EVs employ the Binary Algorithm. The parameters are chosen as $s = 3$ and $r = 60$. Figures 7.3 and 7.5 depict the aggregated power allocated to the EVs vs. the available power C_p, for the constant and dynamic setting, respectively. The allocated power was filtered over a short time scale (10 min) to improve the clarity of the figure. As can be observed the sum of charge rates is always close to the available power. Figures 7.4 and 7.6 depict the average charge rate of 3 randomly selected agents. Even though the agents are only turning on and off at times the average charging rate is equivalent among agents showing a fair share of the allowed on-times. Note that we here can investigate the charge rate instead of the on time due to the fact that the power consumption when on is equal for all agents.

7.5.2   AIMD in a Domestic Scenario

The following parameters are chosen for all the EVs: $\alpha = 0.04 \, \text{kW} \, \text{s}^{-1}$, $\beta^{(1)} = 0.7$, $\beta^{(2)} = 0.98$, and $\lambda = 0.06$. Figures 7.7 and 7.9 depict the aggregated power allocated to the EVs vs. the available power C_p, for the constant and dynamic setting, respectively. The allocated power was filtered over a short time scale (10 min) to improve the clarity of the figure. As can be observed the sum of charge rates is always close to the available power and is able to follow better than the Binary Algorithm, which is expected due to the extra freedom of the controller. Figures 7.8 and 7.10 depict the average charge rate of 3 randomly selected agents. As desired the individual charge rates of the EVs are equalized.

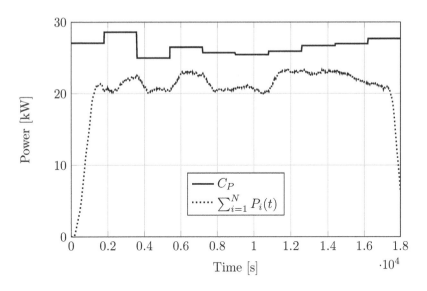

FIGURE 7.3
Constant simulation setting for the Binary Algorithm scenario. The aggregated demanded power vs. the available power C_p.

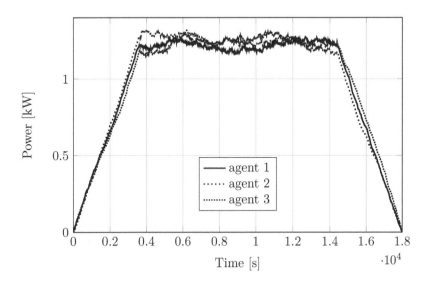

FIGURE 7.4
Constant simulation setting for the Binary Algorithm scenario. The averaged charge rates of 3 randomly selected EVs.

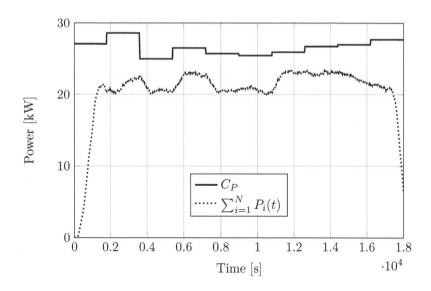

FIGURE 7.5
Dynamic simulation setting for the Binary Algorithm scenario. The aggregated demanded power vs. the available power C_p.

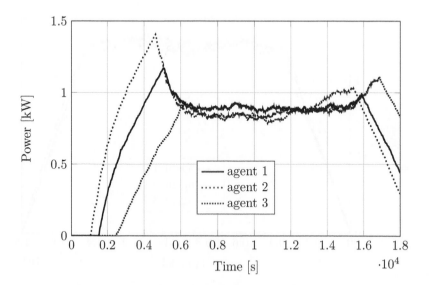

FIGURE 7.6
Dynamic simulation setting for the Binary Algorithm scenario. The averaged charge rates of 3 randomly selected EVs.

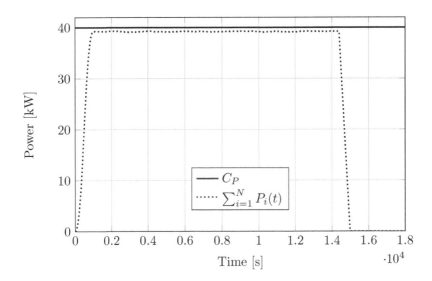

FIGURE 7.7
Constant simulation setting for the domestic AIMD algorithm scenario. The aggregated demanded power vs. the available power C_p.

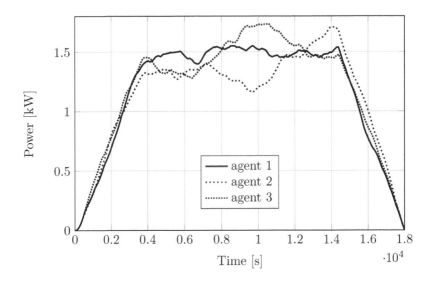

FIGURE 7.8
Constant simulation setting for the domestic AIMD algorithm scenario. The averaged charge rates of 3 randomly selected EVs.

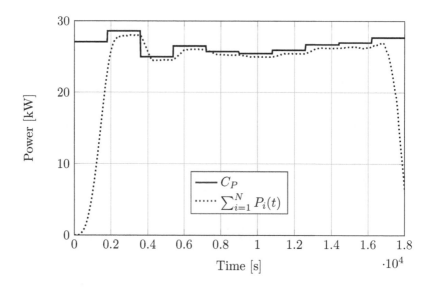

FIGURE 7.9
Dynamic simulation setting for the domestic AIMD algorithm scenario. The aggregated demanded power vs. the available power C_p.

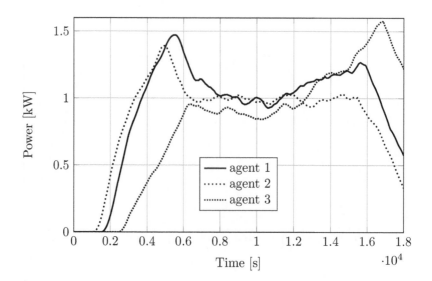

FIGURE 7.10
Dynamic simulation setting for the domestic AIMD algorithm scenario. The averaged charge rates of 3 randomly selected EVs.

7.5.3 AIMD in a Workplace Scenario

The AIMD parameters, apart from α, are selected identical to the domestic scenario: $\beta^{(1)} = 0.7$, $\beta^{(2)} = 0.98$, and $\lambda = 0.06$. α is on the other hand adjusted according to the desired charge rate as proposed in (7.3). Figures 7.11 and 7.12 depict the ratio between the actual and desired charge rate of three randomly selected vehicles, for the constant and dynamic setting, respectively. As can be seen this ratio is equalized as desired. The aggregated power consumption in relation to the available power experiences a similar behavior to the one shown in the domestic setting and is therefore not repeated here.

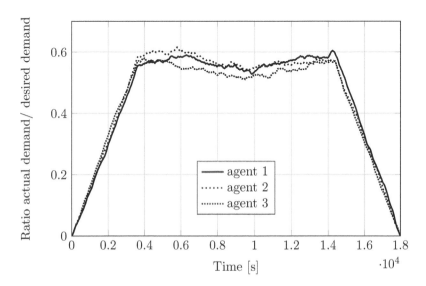

FIGURE 7.11
Constant simulation setting for the workplace AIMD algorithm scenario. The averaged ratio between the charge rates and the desired charge rates of 3 randomly selected EVs.

7.5.4 Binary and AIMD Algorithm Scenario

Here, we investigate the special case where some EVs utilize the Binary Algorithm while others use the AIMD in a domestic scenario. We choose identical parameters as previously used in the simulations. This also means that the EVs do not aim to achieve fairness among all but solely among the ones using the same algorithm. Note that in this regard it is equally possible to control additional groups that aim to achieve another type of fairness among themselves, such as the domestic and the workplace scenario. Note that the resulting aggregated share looks similar as in other scenarios and is therefore omitted.

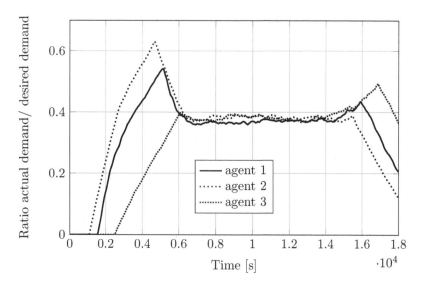

FIGURE 7.12
Dynamic simulation setting for the workplace AIMD algorithm scenario. The averaged ratio between the charge rates and the desired charge rates of 3 randomly selected EVs.

Figures 7.13 and 7.14 depict the average charge rate of 3 randomly selected EVs, where one agent utilizes the Binary Algorithm (EV 1) and two the AIMD algorithm (EVs 2 and 3). As desired the EVs utilizing the AIMD algorithm equalize their charge rate.

7.6   Concluding Remarks

This chapter reviews the main aspects associated with charging of EVs, and discusses issues arising in the design and operation of charging mechanisms. We then discussed two specific charging algorithms for EVs. Inspired by congestion control in communication networks, the resulting EV charging strategies are flexible, efficient and simple to implement.

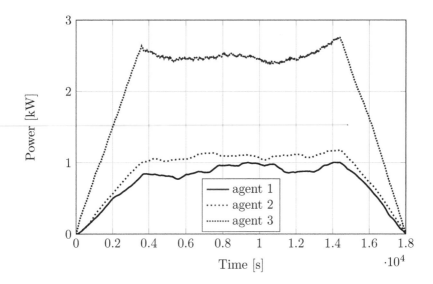

FIGURE 7.13
Constant simulation setting for the binary and AIMD algorithm scenario. The averaged charge rates of 3 randomly selected EVs. EV 1 utilizes the Binary Algorithm while EVs 2 and 3 utilize the AIMD algorithm.

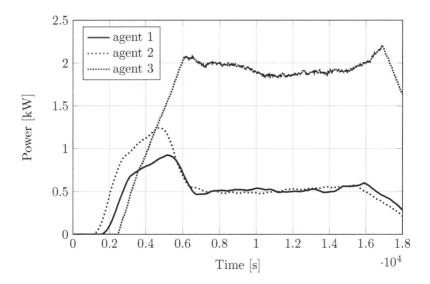

FIGURE 7.14
Dynamic simulation setting for the binary and AIMD algorithm scenario. The averaged charge rates of 3 randomly selected EVs. EV 1 utilizes the Binary Algorithm while EVs 2 and 3 utilize the AIMD algorithm.

FIGURE 7.1?

Constant oscillation setting for the binary and AIMD algorithm scenario. The averaged change value of 5 randomly selected EVs. $\Delta x = 1$ utilizes the binary Algorithm while $\Delta x = 2$ and bigger utilize the AIMD algorithm.

FIGURE 7.?

Dynamic oscillation setting for the binary and AIMD algorithm scenario. The averaged change rate of 5 randomly selected EVs. $\Delta x = 1$ utilizes the binary Algorithm while EVs bigger utilize the AIMD algorithm.

8

Vehicle to Grid

8.1 Introduction

In Chapter 7 we discussed how a fleet of vehicles can be charged when plugged to the grid. During this time however, the grid may occasionally take energy from the connected EVs, either to support power generation at peak moments of the day, or to deliver ancillary services (e.g., frequency regulation). Such an operation is known as V2G (Vehicle-to-Grid), as the power flow now goes from the vehicle to the grid (see for instance [1, 147]). The recent literature contains many examples of research works studying the V2G concept [63], [2], [89]. Issues considered include the ability of V2G to balance the demands of the grid with available supply, the cost returns of V2G operations, and the integration of renewable energy into the V2G concept. In particular, distributed [100] and decentralized [197] mechanisms have been proposed to coordinate EVs to offer V2G regulation services.

While there is a rich literature on how EVs can be used to provide V2G operations, the same level of attention has not been paid to the possible impact of such services. In particular, a discharging of EVs may cause some inconvenience to the owners. For example, V2G operations may affect the ability of the EV user to make the next trip without charging the vehicle, or may even affect negatively the environment. Finally, frequent charging and discharging of the battery associated with V2G operations may also have an impact on the efficiency and on the lifetime of the battery [62].

This chapter is devoted to describe V2G services that can be provided by EVs to the power grid. In particular, we show how to the AIMD-based algorithm illustrated in Chapter 7 can be extended to include V2G operations and active/reactive power exchange. Also, we discuss some (unintended) issues that may arise with V2G operations. In particular, a key conclusion is that treating a fleet of electric vehicles as a virtual storage system is not straightforward, due to the fact that the carbon footprint depends critically on the manner in which energy is drawn from the vehicles.

8.2 V2G and G2V Management of EVs

We now describe how the AIMD-based charging algorithm introduced in Chapter 7 can be easily extended to further include V2G operations, and reactive power exchange. In doing so, we follow reference [173], and we continue to use the notation introduced in Chapter 7.

8.2.1 Assumptions and Constraints

- First, we assume that the charge rate $P_i > 0$ of the ith EV (i.e., the active power drawn from the grid), is continuously adjustable. For simplicity, we also neglect the losses arising from conversion in the battery.

- Second, we denote, by a negative value, the power flowing from the ith EV to the grid, i.e., $P_i < 0$. Accordingly, we implicitly assume that charging circuits should include active rectifiers to perform V2G operations.

- Third, we assume that the ith EV is also able to exchange reactive power Q_i with the grid, to provide ancillary services for reactive power management. We refer to this ability as reactive Vehicle-to-Grid (rV2G) or reactive Grid-to-Vehicle (rG2V). From a hardware perspective, this is also achievable with standard active rectifier connections. We assume that the positive direction of the reactive power flow is from the grid to the EV, in accordance with the sign of the active power flow.

- Fourth, we assume that the apparent power $(S_i = \sqrt{P_i^2 + Q_i^2})$ per EV is limited by an upper bound $\bar{S}_i$ (see for example [31]). Such a constraint is determined by physical limitations such as a maximum inverter current limit.

- Fifth, we assume that the infrastructure is aware of the instantaneous active power demand, the instantaneous available power, and the required and the supplied reactive power. Such knowledge is required to communicate to the EVs when congestion events occur, and also when the charging mode has to switch from V2G to G2V and vice versa.

- We finally assume that the infrastructure is able to send capacity notifications to the EVs. This can be achieved, for instance, by adopting network communication methods, like Wi-Fi, power line communication (PLC), mobile Internet or Zig-Bee. We also assume that the infrastructure communicates to the vehicles when the vehicle mode G2V has to switch to V2G (and vice versa).

8.2.2 Management of Active/Reactive Power Exchange

In Chapter 7, we introduced the maximum (active) power $\overline{P}_i(k)$ that could be supplied to the ith vehicle at the kth time step. In the present context, we must now include upper bounds on the maximum reactive power that can be exchanged with the grid

$$Q_i(k) \leq \overline{Q}_i(k),\tag{8.1}$$

and on the maximum apparent power

$$S_i(k) = \sqrt{P_i(k)^2 + Q_i(k)^2} \leq \overline{S}_i(k).\tag{8.2}$$

In principle, it is desirable that the charge rate $P_i(k)$ is as close as possible to the rated power output of the charger $\overline{P}_i(k)$, as this would imply the quickest (allowed) charge of the vehicle. Also, it is desirable to have the exchanged reactive power as close as possible to the reactive power desired by the grid. However, due to the limitations of the charger outlet, it might not be possible to satisfy both requests at the same time. Thus, in the following discussion, we assume that the active power management (i.e., charging) has a higher priority over reactive power exchange (i.e., ancillary services), and that reactive power exchange does not interfere with the charging process. With this assumption, it is possible to generalize the AIMD algorithm described in Section 7.3.3 to further include the reactive power management step, after the charging step. The overall algorithm is illustrated in Figure 8.1, and consists of two AIMD cycles. The first AIMD cycle involves the computation of the charge rate, and is identical to the one described in Section 7.3.3. The second AIMD cycle is used to compute the reactive power that needs to be exchanged with the grid. Accordingly, the reactive power management does not influence the battery charging process, and is transparent to the EV owner.

Remark 8.1. In specific situations, one may be interested in exchanging priorities.

8.2.3 V2G Power Flows

In the discussion this far we have, somewhat tacitly, assumed that the reactive power flow was also unidirectional (i.e., from the grid to the vehicles). However, both active and reactive power may actually flow in the other direction from the vehicle to grid.

To mathematically embed the possibility of discharging of the EVs we shall adopt the notation of negative power exchange (either active or reactive) to indicate that the power flow is actually from the vehicle towards the grid (V2G). In principle, such a possibility is not allowed by the AIMD algorithm

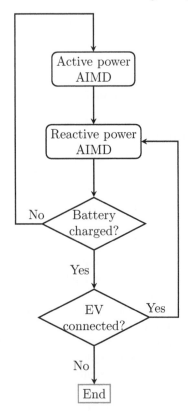

FIGURE 8.1
Flow diagram of the double AIMD algorithm used for active/reactive power
exchange

(positive rates can additionally increase, or multiplicatively decrease, but in
any case they would remain positive rates). For this reason, we assume that
the grid further notifies the EVs whenever the active (or reactive) power flow
is required to change its direction, and we shall call this "mode notification"
(i.e., aG2V rather than aV2G, or rG2V rather than rV2G). When the mode
notification (either for reactive or for active power) indicates V2G mode, then
the vehicles start entering in the additive phase (as in Section 7.3.3) but the
positive additive parameter is now subtracted from $P_i(k)$ and $Q_i(k)$, respectively. When a capacity event occurs (i.e., the power injected by the EVs
exceeds the demand of the grid), the vehicles "decrease" (in absolute value)
the power injected by the usual multiplicative parameter $\beta^{(1)}$ and $\beta^{(2)}$ that
are selected in a stochastic manner.

The active power AIMD in V2G mode (i.e., aV2G) contains the changes
for reverse power flow and is illustrated in Figure 8.2. On the other hand the
active power AIMD in G2V mode (i.e., aG2V) remains unchanged from the

one previously presented in Figure 7.2 besides that the mode notification is processed before the active capacity event. These two modes together describe the active power AIMD shown in Figure 8.1.

The reactive power AIMD similarly has two modes depending on the power flow. Its operation is as in Figures 8.2 and 7.2, when the active power $P_i(k)$ is replaced by the reactive power $Q_i(k)$ and the bound $\overline{P}_i$ is replaced by $\sqrt{\overline{S}_i^2 - P_i(k)^2}$, where $P_i(k)$ is determined by the active power AIMD executed beforehand.

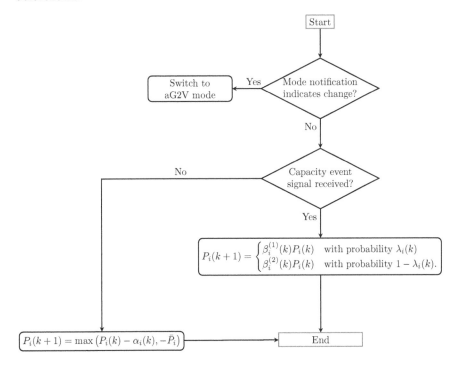

FIGURE 8.2
Flow diagram of the active AIMD algorithm used for V2G charging in aV2G mode

The modified AIMD algorithm (i.e., the double AIMD algorithm with a further mode notification event) can operate in 4 different possible ways, depending on whether the active and the reactive power exchanges actually occur in the G2V or in the V2G direction (i.e., aV2G-rV2G, aV2G-rG2V, aG2V-rV2G, aG2V-rG2V). This means for example when operating in aV2G-rV2G, the EV first process the active AIMD in V2G mode, as illustrated in Figure 8.2 and then consecutively executes the reactive AIMD in V2G mode. A number of different optimization problems can also be implemented using the algorithm. More details can be found in [173].

8.3   Unintended Consequences of V2G Operations

In this section we discuss some consequences, sometimes unintended, of V2G services. In particular, when drawing energy from EVs, three key points need to be addressed: (i) revenues for the owners (e.g., price), (ii) the inconvenience for the vehicle owners, and (iii) the effects on the environment. While most of the literature is focused on the first two points, and is trying to evaluate and quantify the actual inconvenience for the owners (see for instance [62] where the degradation cost of the battery is explicitly taken into account when V2G operations are planned), fewer authors have analyzed the last point, and here we further explore this aspect.

We evaluate the effects on the environment of V2G operations in a simple way.

- We assume that the grid requires a given number of kWh to improve its operation;

- We assume that a fleet of vehicles (including PHEVs and FEVs) are connected to the grid, and are ideal candidates to provide the required energy to the grid. We further assume that the energy stored in the batteries of the vehicles is greater than the energy required by the grid: thus, the grid has some flexibility to choose from what vehicles, and to what extent, it can take the required energy;

- We further assume that the grid can also take the required energy from some connected (conventional) power plants, that can increase their generated power, as an alternative to take energy from vehicles;

- Finally, we wish to find the optimal mix of energy (some from PHEVs, some from FEVs, and some from the power plants) in order to minimize pollution (here, an aggregated mixture of CO, NOx, SOx, and volatile organic compounds (VOCs), as from [75]).

8.3.1   Utility Functions

We solve the optimization problem by formulating utility functions that relate the quantity of energy that is taken from a vehicle to the environmental impact of a vehicle losing this amount of energy. In particular, note that such utility functions are assumed to be increasing with the quantity of energy taken from the vehicle (if a very small quantity of energy is taken from the vehicle, then there are no repercussions on pollution; but the more energy is taken, the more inconvenience is caused to the owner who may have to look for alternative (more polluting) means of transportation). Also, for simplicity, we shall consider piecewise linear functions, although other (more complicated or realistic) shapes of utility functions may be considered as well.

Below we list the main factors that we consider to build the utility functions.

PHEVs: For simplicity, we assume here that PHEVs can either drive in pure EV mode, or can switch to ICE mode. Accordingly, the environmental footprint of a PHEV depends on several factors. Firstly, the distance that a vehicle can drive in full electric mode is a critical factor, and it depends on the energy stored in the battery. If the desired driving distance is larger than the capacity of the PHEV, the vehicle's combustion engine will switch on as soon as the SOC is low. This will have an impact on the environment through the use of carbon based fuels. Therefore, taking electric energy due to V2G operations from PHEVs has the effect of reducing their fully electric mode ranges, and potentially to produce more pollutants. Note that the full electric mode range cannot be computed trivially as it depends upon several factors such as: the state of charge of the battery pack; basic power consumption per kilometer; individual driving behavior; and usage of other electrical appliances (for example, heating, air conditioning, entertainment systems, headlights, or GPS); see [185] or [49]. The driven route also has a strong influence on the available full electric range, as power consumption varies according to driving speed, traffic conditions, and the topology of the terrain, as we have described in Chapter 4. One more subtle factor that should be considered is related to losses caused by energy transfers. For example, continuous charging and discharging could reduce energy efficiency significantly. Also, if too much energy is taken from the vehicle, it may not be able to drive in restricted areas (e.g., in some city centers where only electrically powered vehicles are allowed), and longer journeys may be required (with an associated increase in aggregate pollution production).

FEVs: FEVs are characterized by many of the factors that we have listed for the PHEVs. However, the main difference is that now FEVs cannot rely on an alternative propulsion system (ICE) if not enough energy is available for the next trip. This poses a number of further aspects, that often are not simple to consider and evaluate: (i) *charging*, due to unexpected V2G operations, a vehicle may have to charge during the next trip to reach the final destination, or might remain connected for an extra period before being used for traveling. The emissions due to the extra charging period depend on the generation side; (ii) *second vehicle*, rather than spending time for charging, the owner may decide to take a second vehicle, if available. Accordingly, extra emissions depend on the nominal emissions per kilometer of the second vehicle (or on its state of charge, if it is another FEV or a PHEV); (iii) *public transport*, if a second vehicle is not available, the owner may also decide to take public transport. In principle, such a decision may even be better in terms of pollution, especially if highly developed and environmentally friendly transportation systems are available; (iv) *other measures*, in some cases, there may even not be alternatives, and the next trip may have to be canceled.

Power plants: Obviously, the grid may decide to simply take the required energy from available conventional power plants, rather than from vehicles through V2G operations. For this purpose, one should only consider the power plants that can actually regulate their power output (while we do not consider here the reserves for sudden failing of other generators, or the short-time demand and power matching spinning reserves). Here the utility functions take into account the air pollutants and emissions caused by power plants as a function of the produced energy, and the pollution caused by modulating the power output. Some other factors that need to be taken into account in this case are the emissions related with waste materials (e.g., their disposal), emissions associated with extra fuel and carbon, associated with maintenance, efficiency and losses of the power plants. In particular, note that not all of the previous factors are caused by instantaneous emissions (e.g., maintenance), but still have an impact on the long run, see [186], [75]. For simplicity, we assume that the relationship between the energy delivered by the power plant and the resultant production of pollution is linear. While this relationship is an approximation of the true one [157], it is commonly used in the literature as it represents a good trade-off between simplicity and accuracy; see for instance [75], [15], [139].

8.3.2   Optimization Problem

In the optimization problem, the objective is to provide the required energy, denoted by E_{req} in a region of interest. This energy can be delivered by some connected PHEVs, FEVs, and available power plants. The optimization problem is solved at every time step (where the choice of the time step depends on the specific problem of interest). The optimization objective is to minimize the total pollution caused by FEVs, PHEVs, and power plants. Each of the N participants is assigned a unique index and is assigned to its corresponding group. In particular, let $\mathcal{H}$, $\mathcal{F}$, and $\mathcal{P}$ be the sets of indices that correspond PHEV, FEV, and power plants, respectively. Similarly, let $f_i(\cdot)$, $g_i(\cdot)$ and $h_i(\cdot)$ be the utility functions associated with the ith PHEV, FEV and power plant respectively. Then the optimization problem can be formulated as

$$
\begin{cases}
\min_{E_i} & \displaystyle\sum_{i \in \mathcal{H}} f_i(E_i) + \sum_{i \in \mathcal{F}} g_i(E_i) + \sum_{i \in \mathcal{P}} h_i(E_i) \\[2ex]
\text{subject to} & \displaystyle\sum_{i=1}^{N} E_i = E_{req} & (i \in \mathcal{H} \cup \mathcal{F} \cup \mathcal{P}) \\[2ex]
& -\underline{E}_i \leq E_i \leq \overline{E}_i & (i \in \mathcal{H} \cup \mathcal{F}) \\[2ex]
& 0 \leq E_i \leq \overline{E}_i & (i \in \mathcal{P})
\end{cases}
,
$$

$$(8.3)$$

where the objective is to determine the optimal values of the free variables E_i, that correspond to the energy supplied to the grid by a vehicle (if $i \in \mathcal{H} \cup \mathcal{F}$) or a plant (if $(i \in \mathcal{P})$). Such optimal values are determined by minimizing the sum of produced pollutants (first line), provided that the grid receives the desired amount of energy (second line), and the energies provided by the vehicles (third line) and by the power plants (last line) are feasible. In the case of the vehicles, the energy supplied by the vehicles is constrained between $-\underline{E}_i$ and $\overline{E}_i$, where the lower bound can be negative (if vehicles are actually charged, up to the full level of their batteries), while the upper bound is given by the actual level of energy in their batteries. In the case of the power plants, the upper bounded $\overline{E}_i$ takes into account the maximum energy that can be supplied by the power plant in a time step.

8.3.3 Example

In order to provide some concrete results, we now show an application of the optimization problem (8.3) through a very simple example. In particular, we assume that three vehicles are willing to participate in a V2G energy exchange programme, and that the electricity grid requires 18 kWh. The three vehicles are a PHEV and two FEVs. We assume that the FEV owners will take alternative means of transportation, if required. Batteries and range abilities of the FEVs are those documented for a Nissan Leaf under different environmental conditions ([49]), while data for a Chevrolet Volt is chosen for the PHEV[1]. We assumed that most of the power used to charge the vehicles was generated from renewable sources (data from [75]). Finally, the pollution factor of the power plant was also taken from [75]. Other parameters (e.g., actual SOC, minimum SOC, length of the next trip) were chosen in an arbitrary fashion (see [175] for more details). Then, the results of the optimization problem are reported in Table 8.1. The first two rows of the table correspond to the case where the desired 18 kWh are equally taken from the three available vehicles (no optimization at all). As can be seen, this gives rise to 47.7274 g of pollution (according to the mixture previously discussed). The second two rows show that pollution can be reduced down to about 40 g if the optimization problem (8.3) is solved, and the grid optimally chooses how much energy should be taken from each vehicle. In particular, note that most of the energy is actually taken from the PHEV, as it is more flexible in terms of its propulsion system (i.e., it can also travel in ICE mode). Finally, the last two rows correspond to the case when also the power plant is available. In this case, the overall pollution can further decrease (i.e., below 39 g), and interestingly the optimal solution envisages that one FEV gets actually charged (i.e., operated in the conventional G2V mode) rather than discharged.

---

[1]Manufacturer's Brochure, 2013 Chevrolet Volt, available online at `https://www.vehiclehistory.com/vehicle-fuel-capacity-specifications/chevrolet/volt/2013`. Last Accessed July 2017.

TABLE 8.1
Produced pollution when the required 18 kWh of energy are provided by the available participants

	FEV 1	PHEV 2	PHEV 1	plant 1	Total
E_i [kWh]	6	6	6	–	18
f_i, g_i [g]	18.0389	18.3184	11.3702	–	47.7274
E_i [kWh]	3.0476	3.0755	11.8768	–	18
f_i, g_i [g]	7.4408	10.8901	21.6826	–	40.0135
E_i [kWh]	3.0476	-0.2567	11.9556	3.2535	18
f_i, g_i, h_i [g]	7.4408	2.5134	21.8209	7.1906	38.9658

8.3.4   Alternative Cost Functions

In the previous section we have shown through a simple example that in general it might be more effective to perform V2G operations in a smart manner, rather than equally sharing the V2G services among the available participants. Our optimization problem was formulated in terms of minimum impact on the environment (i.e., by minimizing the quantity of pollutants), but other cost functions can be considered as well. For instance, one may be interested in equalizing the pollution caused by each participant, under the assumption that it would be fairer to ensure that the impact on the environment caused by each participant should be the same. Alternatively, one may be interested in minimizing other functions (e.g., taking the perspective of the grid, it would be interesting to take the energy from the cheapest available vehicles). Similar extensions can be found, for instance, in [174].

8.4   Concluding Remarks

In this chapter we have reviewed the ongoing research on the ancillary services that can be provided by EVs, and on V2G operations in particular. In addition, we have described how a charging algorithm can be easily modified to further account for the possibility to exchange reactive power, and to reverse the flow from the vehicles to the grid. Also, we have shown that poor choices in V2G operations may have severe environmental effects, thereby mitigating one of the principal benefits of plug-in vehicles; namely, that of cleaner air.

Accordingly, an optimal integration of EVs in the power grid is still subject of research, especially with respect to the optimal provision of ancillary services (i.e., reactive power exchange, and V2G).

Part II

The Sharing Economy and EVs

9

Sharing Economy and Electric Vehicles

9.1 Introduction and Setting

Consumer behavior across a spectrum of markets is experiencing a powerful disruptive paradigm shift. Driven by limited natural resources, expanding middle-class, outdated business models, and pressures to reduce wastage, consumers are moving from a sole-ownership, to a shared-ownership model with guaranteed access. Examples of such systems can already be widely found in the realm of mobility services. While the value of the sharing economy is not in question, remarkably, structured platforms to support the deployment and the design of such services, are poorly developed with significant opportunities for improvement. Roughly speaking, several different types of shared products can be discerned.

A. **Opportunistic sharing:** Services based on opportunistic sharing of resources exploit large scale availability of unused resources, and/or outdated business models. Here supply-demand balance is achieved by trading excess supply between actors in return for monetary reward. Sometimes, services of this kind exploit information exchange and brokerage between users to create services; for example, in the trading of flat rate services (i.e., trading of mobile phone minutes). Examples of products in this area include Parkatmyhouse.com[1] and peer-to-peer car sharing services. Trading of mobile phone minutes, unused water allowances, and excess energy are further examples of opportunistic sharing. The key enablers for such products are mechanisms for informing actors of available resources, their delivery, and for their payment.

B. **Federated negotiation and sharing:** Here supply-demand balance is achieved by aggregating actors to empower them to better negotiate with suppliers. Roughly speaking, in this case, groups of users come together to negotiate better contracts with utilities (gas, electricity, water, health), or to provide mutually beneficial services such as collaborative storage of energy. The key enablers for such products are mechanisms for grouping

---

[1]https://www.justpark.com/. Last Accessed July 2017.

together communities and for enforcing contractual obligations for federations of like-minded consumers.

C. **Bespoke shared products:** Driven by the success of existing shared services, the rapid development of Internet of Things (IoT) technologies, and the need for sustainable design, a number of large companies are already exploring ways to design products with the specific objective of being shared.

D. **Shared products created by policy gradients:** Significant opportunities are also arising in response to gradients resulting from national policy. In many areas, opportunities exist to monetize the trading of transferable credits and risk that arises as a result of government policies. The key enablers for such products are analytics to ensure fairness in allocation of risk and credits.

9.2 Contributions

Clearly, EVs and the Sharing Economy are a very good match and our objective in this part of the book is to describe some of the work that occurs at the interface of these two domains. While many companies are already active in this space, including BMW, Daimler, Fiat, Peugeot, Volkswagen, and Renault, much work remains to be done. For example, as we write this text, generic cloud based brokerage engines to both host and support the development of mobile and ad-hoc collaborative consumption services are needed by both consumers and application developers. Also, analytics and tools to help design collaborative consumption services do not currently exist to any significant degree. Finally, general platforms for bringing together communities, and aggregating and federating informed consumers, in a manner that addresses the contractual, liabilities and privacy aspects of users, to create virtual super-users for the purpose of negotiation, do not exist in the area of shared trading. While each of these gives rise to opportunities for research in several areas, our focus in this part of the book is on the development of bespoke analytics and tools to help design collaborative consumption services in the context of EVs. Specifically, we are interested in alleviating certain forms of *angst* associated with EV ownership. In the first chapter of this section we deal with range anxiety, and in the second chapter with charge point anxiety.

Notes and References

This part is based on the following papers by the authors and their co-workers. The chapter on car sharing is based on papers written in collaboration with Wynita Griggs, Christopher King, Paul Borrel, Mingming Liu, and Karl Quinn [106], [121][2]. The chapter on charge point access is based on the work [76] that was carried out with the collaboration of Wynita Griggs, Jia Yuan Yu and Florian Häusler[3], while the design of the *Charge Point Adaptor* (CPA) was done in collaboration with Rodrigo Ordóñez-Hurtado, with some input from Brian Mulkeen and Eoin Thompson.

---

[2]©IEEE. Figure 10.9 and Table 10.6 reprinted, with permission, from [87].
[3]©IEEE. Reprinted, with permission, from [76].

10

On-Demand Access and Shared Vehicles

10.1 Introduction

We have already discussed some of the issues that impede the adoption of electric vehicles. In this chapter, we focus on the most pressing of these issues, **range anxiety** (see below), and to a lesser extent, **vehicle size** by developing a flexible vehicle access model to alleviate both of these issues. Specifically, a solution is proposed to some of these problems based on car sharing. This idea is an embodiment of flexible vehicle access that was first suggested in [83] and further developed and analyzed in [107]. Indeed, the timeliness of the idea is evidenced by the fact that several automotive manufacturers, including Fiat, and Volkswagen, have developed a form of car sharing with similarities in both goals and implementation, to that described here. The principal difference to these embodiments is that we advocate and describe an **on-demand model**, whereas car manufacturers are suggesting a model that guarantees access for a fixed, but small, number of days. Our contribution in this chapter is thus threefold. We develop tools to design an on-demand service using ideas from queueing theory and using predictive analytics. We then demonstrate, in the context of real Irish mobility patterns, that such an on-demand service is economical, both in terms of the number of ICE vehicles needed, and in terms of real additional cost to vehicle manufacturers. Finally, we provide some quantitative analysis of how much fleet emissions can be reduced, even by introducing a very small number of EVs.

We note that in our models it is assumed that extensive trips are planned and the on-demand scheme responds to demand announced on the previous day (or several days in advance). There are, of course, other possible on-demand scenarios. The assumptions described below pose no fundamental restrictions in this respect and the methods presented here can be extended. The important point to note in this context is that the customer is not assigned a fixed number of days per year in which access is guaranteed; rather the user is allowed to request a vehicle at any time. This is the meaning of *on-demand* in our context.

10.2   On Types of Range Anxiety

Our basic assumption is that the range anxiety problem and vehicle cost are
the major barriers to the purchase of electric vehicles. For the purpose of this
chapter we consider the term range anxiety to mean the *angst* of a vehicle
owner that he or she will not have enough range to reach the destination
without the need for recharging. Roughly speaking, one may consider two
types of problems associated with range anxiety.

(i) The first problem is associated with an inadequate battery level to complete a trip while driving to a destination.

(ii) The second problem is associated with electric vehicles being unsuited to
the trip distribution demanded by the user. For example, a long trip, or
vacation (or indeed a trip where a large luggage load is required), are all
problematic for a typical electric vehicle.

In many cities the first of these problems can to a large extent be avoided
with adequate trip planning. For example, in cities where single dwelling
households with garages or driveways are common, it is possible to charge
vehicles adequately overnight to have a full battery charge the next morning.
As we shall see later, this is, in most cases, adequate for the majority of trips.
Since most UK and Irish cities are of this type, we shall assume that overnight
charging is possible and focus on the second of the above range anxiety issues.

It is worth noting that the issue of range anxiety has been the subject of
the attention of policy bodies and car manufacturers over the past number
of years. Roughly speaking, research attention has focused on three areas: (i)
better batteries; (ii) optimizing energy management in the vehicles; and (iii)
novel energy delivery strategies for electric vehicles. Big efforts on designing
better batteries have been made worldwide [14]. To optimize energy management, manufacturers have looked both within the electric vehicle through the
management of the vehicle sub-systems, and outside of the vehicle through
the use of energy-aware routing and the use of special lanes (see the eCo-
FEV project[1]). Finally, several methods of delivering energy to the vehicle
have been suggested. These include battery swapping, as was advocated by
BetterPlace[2], under-road induction, and fast charging outlets. These activities have addressed for the most part item (i) above; and largely ignore the
inconvenience associated with item (ii).

---

[1]http://www.egvi.eu/projectslist/23/37/eCo-FEV. Last Accessed July 2017.
[2]https://web.archive.org/web/20080907071156/http://www.betterplace.com/. Last
Accessed July 2017.

10.3   Problem Statement

Our car sharing concept closely follows the flexible access suggestion in [83] in the following manner.

When an electric vehicle is purchased, the new EV owner also automatically becomes a member of a car sharing scheme, where a shared vehicle may be borrowed from a common pool on a 24 h basis. The shared vehicles are large ICE-based vehicles suitable for long range travel and with large goods transportation capacity.

Remark 10.1. We suggest free membership of the scheme, but a pricing model could be implemented to regulate demand on weekends, public holidays, or other occasions when synchronized (correlated) demand is likely to emerge, or to regulate emissions. Further, if the shared ICE-based vehicles are chosen to be sufficiently high-end, then a further incentive for consumers to purchase electric vehicles is provided.

A number of issues need to be resolved before any such system could be deployed. These issues reduce to the marginal cost of the system. More specifically, we wish to determine if such a sharing concept could be deployed giving reasonable QoS to the electric vehicle owner, without significantly increasing the cost of each vehicle. Referring to Figure 10.1, this amounts to asking whether a reasonable QoS can be delivered when M, the number of shared ICE-based vehicles, is significantly less than N, the number of purchased EVs. To answer this question, we consider two scenarios.

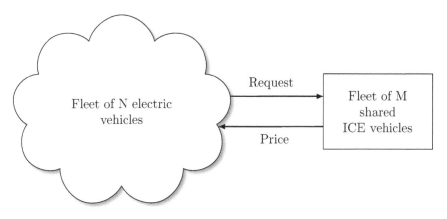

FIGURE 10.1
Car sharing concept

1 **Spontaneous journeys :** If we want to achieve a low probability for the event that an individual requesting a vehicle is not allocated one, how should the proportion $\frac{M}{N}$ be chosen?

2 **Planned journeys :** For fixed M and N, how many days in advance does a user have to make a reservation so that the probability that the request is declined is lower than some very small constant?

In solving the above problems we shall make the following assumptions.

I. Membership assumptions

I.1 A shared vehicle is borrowed for a single "day" period and returned to the pool of shared vehicles to be shared again the next day. In other words, at the beginning of each "day", M cars are available for sharing.

I.2 Customers collect their vehicles from a number of sharing sites. They either park their own EV at the site when collecting the ICE vehicle, or travel there using some other form of transport.

I.3 Members of our car sharing scheme do not have access to an ICE based vehicle other than through the car sharing scheme.

I.4 Members of the car sharing community are willing to accept the same QoS metric for vehicle access.

II. Demographic assumptions

II.1 Long journeys in private cars are rare, meaning that the range of an electric vehicle, even under worst case conditions (e.g.: air conditioning use, traffic congestion, bad weather), should be sufficient for most journeys.

II.2 Most urban dwellings are houses rather than apartments, meaning that there are no structural impediments to overnight charging, and that a full overnight charge should be sufficient to satisfy the needs of most daily mobility patterns.

The validity of these latter assumptions is the subject of the next section.

10.3.1   Data Analysis and Plausibility of Assumptions

Assumption I.1 is based on statistics of mobility patterns described below, which show that most drivers require their car for a whole day, if it is required. Assumption I.2 describes a common practice in car sharing schemes, while I.3 is a simplifying assumption.

To make a case for the plausibility of Assumptions II.1 and II.2 we examined publicly available data on contemporary Irish mobility patterns.

Data for the creation of Table 10.1 and Figures 10.2, 10.3 and 10.4 were obtained from the 2009 Irish National Travel Survey (NTS) Microdata File, Central Statistics Office, ©Government of Ireland[3]. In the NTS, respondents were asked to provide details about their travel for a given (randomly selected) 24 h period, which roughly corresponded to a day of the week.

Table 10.1 shows the percentages of people who drove private vehicles (over the 24 h period that they were queried about) for cumulative daily distances greater than 50 km, 75 km and 100 km. Figure 10.2 relates to those people who were questioned about their travel over the 24 h "Monday" period (i.e.: row two of Table 10.1), and depicts number of people versus total distances they drove in private vehicles over that 24 h Monday period. Figure 10.2 illustrates a trend observed in the percentages in Table 10.1; namely, that longer cumulative journeys over the course of a day were rare[4]. For respondents who drove cumulative distances greater than 75 km over a 24 h period (see the third column of Table 10.1), Figure 10.3 illustrates those hours of a 24 h period over which respondents had their vehicles in use (many vehicles were in use roughly between 8 am and 6 pm), and Figure 10.4 depicts number of respondents versus total time (out of a 24 h period) their vehicle was in use.

TABLE 10.1

Percentages of people who drove cumulative distances greater than 50 km, 75 km and 100 km over a 24 h period

Sample Population	50 km	75 km	100 km
Monday	23%	12%	7%
Tuesday	23%	14%	8%
Wednesday	23%	14%	7%
Thursday	26%	18%	11%
Friday	26%	17%	9%
Saturday	24%	15%	9%
Sunday	24%	17%	11%

---

[3] Central Statistics Office, National Travel Survey 2009, available online at https://www.ucd.ie/t4cms/NTS%20Report%202009.pdf. Last Accessed July 2017; and National Travel Survey 2009 Codebook for Anonymised Microdata Files, available online at http://www.ucd.ie/issda/static/documentation/cso/nts-2009.pdf. Last Accessed July 2017.

[4] Graphs of the nature of Figure 10.2, but concerning travel over the other days of the week, were similar in shape to Figure 10.2, and have thus been omitted.

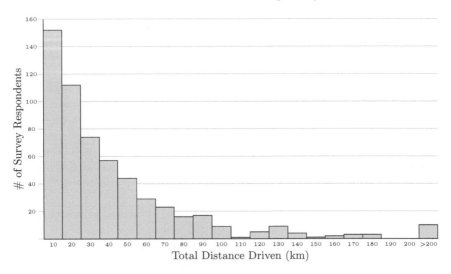

FIGURE 10.2
Number of survey respondents reporting about their travel for the 24 h "Monday" period, versus total distances they drove over that period

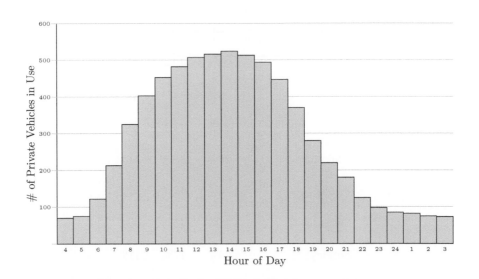

FIGURE 10.3
Number of respondents (who drove >75 km total daily distance) using their vehicles, versus hour of the day

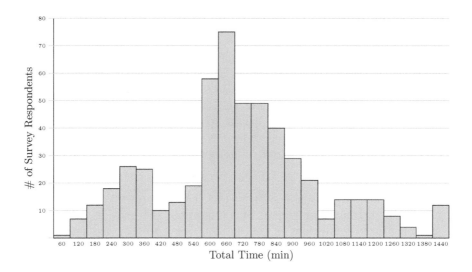

FIGURE 10.4
Number of respondents (who drove >75 km total daily distance), versus total
time that their vehicle was in use over their 24 h survey period

10.3.2 Comments on NTS Dataset

The NTS dataset allows us to make some very general statements which support the assumptions listed above. As might be anticipated, longer journeys are not as frequent as shorter ones, and vehicles do indeed tend to be in use for a single day for each user. Thus, the dataset appears to be indeed consistent with the Assumptions II.1 and II.2 above. However, the dataset is subject to certain limitations due to its size and the manner in which it was collected, and it is important to comment on these before proceeding.

- First, the dataset does not take into account correlated driving behaviors between days. However, we assume that drivers making regular long-distance trips would not purchase EVs. Consequently, the tail of the distributions following from the datasets should be representative of the qualitative behavior of drivers making regular short trips, but who infrequently need larger vehicles or vehicles for larger trips. In what follows we shall assume that the typical EV owner follows a distribution given by these datasets although clearly from the above discussion this is a simplification as driving behavior varies across EV owners.

- Second, it is possible that the car sharing scheme may affect the driving patterns of scheme members (EV owners). For example, given ICE vehicle access, long-distance driving may become more attractive. In the analysis to follow (in the next section), this feedback mechanism is also ignored to keep matters simple.

- The data also indicates that vehicles are "in use" for the entire working day. We assume that vehicles are charged overnight to full capacity and that it is not necessary for a vehicle to be again charged during the day. As we have already mentioned, this latter assumption is consistent with cities such as Dublin where garage space or a driveway is the norm, and is consistent with charging times for standard home charge points. The interested reader is referred to [46] for further information about driver charging behavior.

Given the above discussion, it is clear that the presented data provides a plausibility argument in support of Assumptions II.1 and II.2 subject to the aforementioned limitations of the dataset. For a further and more detailed discussion of the limits of using such datasets the interested reader is referred to [47].

10.4   Mathematical Models

Using elementary probability and queueing theory methods, we now pose solutions to the QoS problems presented in Section 10.3. Consider a population of N electric vehicle owners (i.e.: N "users") who occasionally require access to an ICE-based vehicle (ICE or Internal Combustion Engine Vehicle (ICEV)) for a non-standard trip (either a long-range trip or a trip where large load carrying capacity is required). We assume that a user will keep the ICEV for a full day, based on the driver behaviors described in the previous section. Thus, each day is characterized by the number of users who require an ICEV on that day. There is a fleet of M ICEVs available to satisfy this need. The main question is then to determine the relation between M and N. This will be determined by requiring some QoS conditions to be met. For example, a QoS condition might be a guarantee on the probability of finding an ICEV available.

An important assumption in what follows is that all users are willing to accept the same QoS metric for vehicle access. This is defined by a probability of a scheme member not being able to access a vehicle when required. Similar assumptions are standard in the networking community, and are based on the assumption that a member of the car sharing scheme will not join the scheme unless the given QoS metric is acceptable. Clearly, different levels of service can be provided if groups of users are willing to pay for a higher level of service. However, to keep matters simple here we assume that drivers who sign up for the scheme are prepared to be served using the same QoS metric.

10.4.1 Model 1: Binomial Distribution

In the simplest model, each user independently requests an ICEV each day with probability p. Based on the data available in Table 10.1 we may estimate this probability as $p = 0.0886$, assuming that ICEVs are used for trips over 100 km. The assumption that cars are required for a whole day are justified by the data in Figures 10.2 to 10.4, which show that the overwhelming majority of cumulative trips over 75 km require a period of 5 h or more. With the available data we are not able to estimate the proportion of drivers that make long trips on a regular basis, who would not participate in the scheme. On the other hand it may be expected that some ICEVs are requested even though the actual trip length turns out to be less than 100 km. In the following we will use $p = 0.1$ to illustrate the applicability of the proposed scheme.

Thus, the number of requests X each day is a binomial random variable:

$$X \sim \text{Bin}(N, p).$$

The mean number of requests per day is Np, and the standard deviation is $\sqrt{Np(1-p)}$. In principle, the number of requests may be anything from 0 to N, but for large N it is very unlikely that X will deviate from the mean by more than a few standard deviations.

The QoS condition we consider here can be quantified as follows. For each $M \leq N$, define

$$Q(M) = \mathbb{P}(X > M).$$

Then the QoS condition could be to find the smallest M such that $Q(M) < \epsilon$ for some specified $\epsilon > 0$. For any given N and p this can be calculated explicitly using the formula

$$Q(M) = \sum_{k=M+1}^{N} \binom{N}{k} p^k (1-p)^{N-k}.$$

However it is more useful to get an approximate formula from which the scaling relation can be read off. For N large enough we can use the normal approximation for the binomial, which says that

$$Z = \frac{X - Np}{\sqrt{Np(1-p)}}$$

is approximately a standard normal random variable. There is a standard rule of thumb regarding applicability of the normal approximation for the binomial, namely that

$$N \geq 9 \max\left(\frac{p}{1-p}, \frac{1-p}{p}\right).$$

Using the normal approximation, we have

$$Q(M) \simeq \frac{1}{\sqrt{2\pi}} \int_r^\infty e^{-\frac{1}{2}X^2}\, dX, \qquad r = \frac{M - Np}{\sqrt{Np(1-p)}}.$$

This readily yields estimates for M in order to satisfy a desired QoS condition. For example, in order to satisfy the QoS condition

$$Q(M) < 0.05$$

meaning a less than 5% chance of not finding an ICEV available, it is sufficient to take

$$r \geq 1.65 \iff M \geq Np + 1.65\sqrt{Np(1-p)}.$$

For example, using the values $N = 1000$ users and $p = 0.1$ for the probability of a user requesting an ICEV, this provides a value $M \geq 116$.

10.4.2   Model 2: A Queueing Model

When the number of requests exceeds the number of available ICEVs, a queue forms and users must wait one or several days until a vehicle becomes available. It is desirable to keep the probability of long delays small, and this can be achieved by appropriate scaling of M with N. This is the subject of this section.

Let X_n be the number of outstanding requests at the end of the nth day, and let A_n be the number of new requests that arrive during the nth day. Since the number of vehicles is M, the relation between these variables on successive days is

$$X_{n+1} = \max\{0, X_n - M\} + A_{n+1}.$$

That is, the queue length is reduced by M at the start of each day (but not reduced below zero), and is then increased during the day by the number of new requests.

The QoS condition is to ensure that X_n is unlikely to be large, implying that users are unlikely to have to wait a long time before being assigned an ICEV. Since M users can be serviced each day, in the worst-case a user must wait $\lfloor X_n/M \rfloor$ extra days until service, where $\lfloor x \rfloor$ gives the largest integer less than or equal to x. We consider the QoS condition which guarantees that the probability that any user needs to wait d_w extra days or more is less than ϵ, that is

$$\mathbb{P}(X_n > d_w M) < \epsilon.$$

By choosing M sufficiently large we can guarantee that this probability is small. The following lemma provides bounds for the probability of large queues. These bounds provide easy estimates which can be used for the choice of M.

Lemma 10.2. *Define*

$$\mu = M - Np, \qquad \sigma^2 = Np(1-p), \qquad \gamma = \frac{\mu}{\sigma^2}.$$

Then for all $k \geq 1$,

$$\mathbb{P}(X_n > d_w M) \leq \frac{1}{2} e^{-(d_w-1)M\gamma} \left(e^{\mu\gamma/2} - 1\right)^{-1}.$$

Proof. A proof of the Lemma 10.2 can be found in [106]. □

Using the bound in Lemma 10.2 we find a sufficient condition to guarantee the QoS bound, namely

$$\frac{1}{2} e^{-(d_w-1)M\gamma} \left(e^{\mu\gamma/2} - 1\right)^{-1} < \epsilon.$$

For a given d_w and ϵ we may use this to find a value for M needed to meet the QoS condition. For example, using the same values as above $N = 1000$, $p = 0.1$, $\epsilon = 0.05$, and taking $d_w = 3$ (meaning that the probability that there is a customer who waits more than 4 days is less than 5%), we find that the inequality is satisfied whenever $M \geq 103$. Taking $d_w = 2$ we find $M \geq 105$, and with $d_w = 1$ we find $M \geq 121$. For a fleet size of 20 000 vehicles with $M = 2000$ (10%) and $M = 3500$ (17%), the behavior of the bound is depicted in Figures 10.5 and 10.6, respectively, for various estimates of the probability of a long distance trip. As can be seen, the bound tends rapidly to zero, indicating that the probability of waiting for a shared vehicle longer than one or two days vanishes rapidly.

10.4.3   Two Opportunities for Control Theory

Before proceeding we note briefly that the aforementioned dimensioning problems offer opportunities for control engineers. These stem from two basic flaws in our model.

(i) **Synchronized demand:** First, the statistical model does not take into account periods of synchronized demand. For example, during periods of vacation, hot weather, etc., demand profiles are almost certainly going to be very different from the statistics we have presented, and are likely to be highly synchronized. To address this problem, two avenues of actuation are possible. One could simply make the number of shared vehicles increase with demand. An alternative is to include a price signal to moderate demand during periods of synchronized requests. This is depicted in Figure 10.7 where the strategy tries to create a pricing structure that maintains the QoS at some desired level during periods of synchronized demand. The design of such a pricing signal (integral action) is non-trivial due to all the

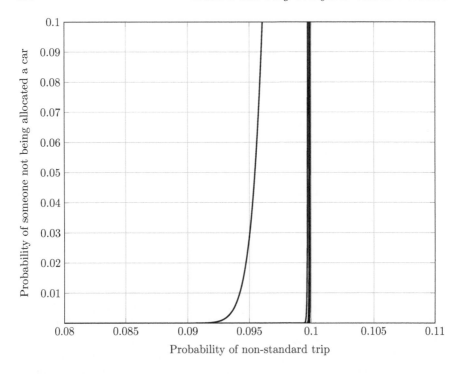

FIGURE 10.5
Probability (bound) of not finding a car with N = 20 000 and M = 2000 (10%).
$d_w = 1$ is the left most curve; $d_w = 6$ is the right most curve. Note that curves
for $d_w = 2, ..., 6$ are hardly distinguishable

usual reasons; feedback delays in the control loop; the need for special infrastructure, and the use of a single signal for all users. In addition, a more
serious concern is that the objective of the control is to smooth demand
and break up synchronization; rather than to just attenuate synchronized
behavior. A further complication is that the signaling should also ensure
a degree of fairness of access for subscribers to the scheme, while at the
same time preserving the privacy of individual users.

(ii) **Prediction and optimization under feedback:** Another problem is
that our set-up assumes decoupling between the statistical model and the
car sharing scheme. Recall that our system was designed by first building a
statistical model, and then using this model to dimension the car sharing
scheme. This set-up ignores the effect of the car sharing scheme on the
statistics governing the model. Put simply, the availability of a free car
is quite likely to influence the likelihood that users will avail of a car
sharing service. A more advanced design would include a model of user
behavior and dimension the number of free cars in a closed loop fashion.

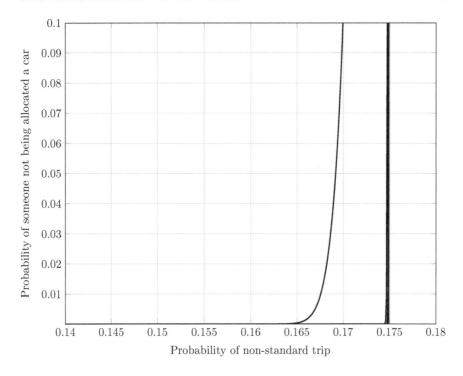

FIGURE 10.6
Probability (bound) of not finding a car with N = 20 000 and M = 3500 (17%).
$d_w = 1$ is the left most curve; $d_w = 6$ is the right most curve. Note that curves
for $d_w = 2, ..., 6$ are hardly distinguishable.

This is clearly a big topic (prediction and optimization under feedback)
with links to areas such as *reinforcement learning* and *adversarial game
theory*, but it is beyond the scope of the present book.

10.5 Financial Calculations

We now explore the financial model associated with this car sharing model.
Initially, we will model our calculations based on a shared fleet comprised
solely of a single ICE vehicle type: the Volkswagen (VW) Golf class car. In
this model we are essentially comparing the similarly sized VW Golf and the
Nissan Leaf to focus solely on examining range anxiety of the kind discussed
above.

Subsequently, we will model our calculations based on a shared fleet that is
constructed to reflect the broader needs of the EV owners; namely, sometimes

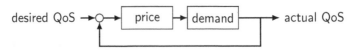

FIGURE 10.7
Schematic of pricing strategy

a family car is needed, sometimes a large vehicle is needed for transporting goods, and sometimes a smaller car is required for short out of town trips by a single person or couple. In this subsequent model we focus on both range anxiety and a range of vehicle sizes. To do this we make some simplifying assumptions. Instead of modeling the demand for various vehicle types, which is the correct method of analysis, we shall weight various vehicle classes to construct a vehicle fleet. We assume that these weights reflect demand. This simplifying model is adopted for two reasons; to keep analysis simple, and to reflect the fact that data of the type needed to build a multi-dimensional queueing model is not available to us.

In both model calculations we use VW vehicles. Note that VW is a relatively high-end marque, and the vehicle fleet could be constructed in a manner that is considerably cheaper. Table 10.2 illustrates the composition and pricing of our shared vehicle fleet (including the single vehicle VW Golf price). Vehicle purchase prices were sourced from the VW website (www.volkswagen.ie) on 23rd September, 2013. The weighted average vehicle purchase price given the fleet make-up is €23,308.

TABLE 10.2
Composition of the shared vehicle fleet

Purpose	% of Fleet	Vehicle Purchase	Price
Singles/Couples	20%	VW Polo	€14,195
Family without luggage	20%	VW Golf	€19,995
Family with luggage	50%	VW Passat	€27,165
Transport	10%	VW Passat Estate	€28,870

We also assume a fleet of N Nissan Leaf electric vehicles. These electric vehicles retailed for €25,990 each in Ireland at the same time (23/9/2013). It is important to note that there are more expensive and less expensive ICEV and EV offerings.

Ahead of detailing the two model calculations we recall Table 10.1. Table 10.1 indicates that the probability of cumulative journeys greater than 75 km is approximately 0.15 (average), and of 100 km is 0.09 (average). If we conservatively assume that the daily range of a fully charged electric vehicle is 75 km, then it follows from Figure 10.6 that most customers will be allocated an ICE-based vehicle within 3 days for $(M, N) = (3500, 20000)$; namely, if $M/N = 0.17$. If we assume that the daily range is 100 km, then it follows from Figure 10.5 that most customers will be allocated an ICE-based vehicle within 3 days for $(M, N) = (2000, 20000)$; namely, if $M/N = 0.10$. In essence, in order to provide 20 000 EVs with an ICEV within 3 days then we need an additional 2000 ICEVs to cater for journeys of 100 km and an additional 3500 ICEVs to cater for journeys of 75 km.

10.5.1 Range Anxiety Model (VW Golf vs. Nissan Leaf)

We utilize a financial model for both the car sharing models, which is in line with current and well known car financing structures. We take an approach that amortizes the purchase of a new ICEV over a three year period with a 20% straight line depreciation Year on Year (YoY). However, first year depreciation is at 40% as suggested as standard by AA[5]. This gives three year depreciation costs for a VW Golf as listed in Table 10.3.

TABLE 10.3
Depreciation costs of the VW Golf fleet

	Start Value	Year 1	Year 2	Year 3
Car Value	€19,995	€11,997	€9,598	€7,678
YoY Depreciation (%)		40%	20%	20%
YoY Depreciation (€)		€7,798	€2,399	€1,920
% Value vs. Start		60%	48%	**38%**
GMFV				**40%**

It is important to note that we also make the assumption that the ICEV is sold in year three for a value close to its depreciated value. Motor companies such as Hyundai, Ford, etc. provide Guaranteed Minimum Future Value (GMFV) mechanisms such that the depreciated value of the car is close to the GMFV of the car after a given time that is typically three years. The three year amortization of depreciation and GMFV allow us to financially model the depreciation costs of the ICEV within a three year bound.

Considering the revenues from 20 000 EVs at a cost of €25,990 is €519.8 million we will now present the additional cost of this car sharing model using the threee year financial approach outlined above.

In the case of 100 km journeys we have described how 2000 ICEVs are required. As we are using a VW Golf in this model then the 2000 ICEVs would have once off costs of €40m, which is 7.7% of the EV revenues. However, such a once off financing mechanism is not typical of financing car purchases in the car industry. Instead, we utilize the three year amortization and GMFV method

---

[5]http://www.theaa.com/car-buying/depreciation. Last Accessed July 2017.

outlined above so as the costs can be modeled more realistically. In Table 10.3 the sum of the single VW Golf depreciation values (YoY Depreciation €) over the three year period is €12,317. In year three we dispose of the VW Golf asset at, or near, the GMFV value of ≈ 40%. As such, the cost of the VW Golf over three years then becomes the depreciation costs of €12,317. As we require 2000 VW Golf cars for this car sharing model this then equates to a fleet cost of €24.6m over three years, or an average of €8.2m per annum. This represents an annual cost overhead of 1.5% per annum for three years for 100 km journeys serviced by a fleet of 2000 VW Golf cars within three days. In the case of 75 km journeys we require 3500 VW Golf cars, which has a similarly calculated annual cost overhead of 2.7%.

10.5.2   Range Anxiety Model with a Range of Vehicle Sizes

This model will utilize the same three year financial approach for a range of ICEVs as used in the VW Golf model. However, the cost of the ICEV is the weighted average, €23,308, of a range of VW ICEV types taken from Table 10.2. In this scenario we are modeling both range anxiety and the ability to cater for ICEV usages for a family (i.e., VW Passat) with luggage and transport purposes (i.e., VW Passat Estate).

TABLE 10.4
Depreciation costs of the weighted ICE fleet

	Start Value	Year 1	Year 2	Year 3
Car Value	€23,380	€13,985	€11,188	€8,950
YoY Depreciation (%)		40%	20%	20%
YoY Depreciation (€)		€9,323	€2,797	€2,238
% Value vs. Start		60%	48%	**38%**
GMFV				**40%**

The revenues from 20 000 EVs at a cost of €25,990 remain unchanged at €519.8 million.

As per the previous model, 100 km journeys require 2000 ICEVs. As we are using a weighted average of different ICEV class cars in this model then the 2000 ICEVs would have once off costs of €46.6m, which is 9% of the EV revenues. In Table 10.4 the sum of the depreciation values (YoY Depreciation €) over the three year period is €14,358. As we again require 2000 ICEVs then this equates to a fleet cost of €28.7m over three years, or an average of €9.6m per annum. This represents an annual cost overhead of 1.8% per annum for three years for 100 km journeys that are serviced by a range of ICEV within three days. In the case of 75 km journeys we require 3500 ICEVs, which has a similarly calculated annual cost overhead of 3.1%.

10.5.3   Financial Assumptions and Key Conclusions

In both models we assume further incremental costs such as traveling to ICEV location, parking logistics, car cleaning, annual servicing, and all other operational costs are either mitigated via an implemented pricing model and/or absorbed by the fixed cost structure already with the dealership, rental company, and so on. We also assume that at the end of the three year period the EV owner must stop using the service or replace that EV with a new EV to start another three year cycle, which maintains the calculations and logic of the approach over terms greater than an initial three years.

Assumptions in the financial model are justified in the use of standard financing models used in car finance, AA derived depreciation models, and weighted average EV and ICEV vehicle purchase prices. However, we do note that our assumptions ignore synchronized demand, are not based on demand distributions for vehicle classes, and mileage restrictions may need to be offset.

In the model where we used VW Golf to focus solely on alleviating range anxiety the overall percentage cost is between 4.5% (100 km, three years) and 8.1% (75 km, three years) of the 20 000 EV costs. In the weighted average model where we focused on alleviating range anxiety, and also offering a range of ICEV size options, then the overall percentage cost is between 5.4% (100 km, three years) and 9.3% (75 km, three years) of the 20 000 EV costs. Note, for further clarification the boundaries of our model ranges from 3.3% overall costs (2000 VW Polo, 100 km) to 11.7% overall costs (3500 VW Passat Estate, 75 km).

To place these figures into context, consider Table 10.5. As can be seen, the cost of the car sharing scheme is considerably less than the average level of subsidy afforded to electric vehicles in major western countries; in 2013 at approximately 23% for a Nissan Leaf EV. Further incentives to encourage the uptake of electric vehicles might include a combination of car sharing and subsidies or replacing the subsidies with an increased level of car sharing. Whether car sharing can really encourage the uptake of such vehicles can only really be tested through implementation. However, we believe that we have demonstrated that a significant range related issue can be solved using this idea, and that this will ultimately affect market growth in a positive manner.

To conclude it is worth recalling some of the assumptions underlying our analysis.

- Our analysis ignores synchronized demand (at weekends or during holiday periods). We argue that a pricing structure could be enforced to break up this demand and give a degree of QoS to the scheme members.

- As we have mentioned, our combined fleet analysis is not based on demand distributions for each vehicle class. Rather, it is a simplified calculation to give the reader an indicative picture of the cost of the scheme.

- We have ignored the fact that the shared vehicles would be in continuous use thereby rendering their value lower than standard GMFV. However,

TABLE 10.5
Subsidies to EV purchase (direct and indirect) and cost of Nissan Leaf. Data sourced from Nissan: September 2013.

Country	Subsidy	Cost (Nissan Leaf)	Percentage
Ireland	€5,000	€25,990	19%
Belgium	€9,000	€29,890	30%
France	€7,000	€30,190	23%
Portugal	€5,000	€31,100	16%
United Kingdom	£5,000	£20,990	25%
United States	$7,500	$28,800	26%

GMFV mileage restrictions may be offset through gains made when bulk buying 2000 or 3500 ICEVs, and/or through negotiating greater mileage restriction caps, and/or by ensuring a reduction in any additional charges beyond the restricted mileage, and so on. Additionally, other depreciation models are easily incorporated into the given framework. Note that depreciation of the entire cost of the shared vehicles over a three year period to zero value is also very cost effective when compared with the average cost of government subsidies (7.7% for a VW Golf in the 100 km model to 15.7% for a weighted average ICEV in the 75 km model, in comparison with the 23% average subsidy). A fleet of 3500 VW Passat Estates in the 75 km model depreciated to zero is more efficient than subsidies (19.4% compared to 23%).

Notwithstanding the above facts, we have shown how the costs from the two model calculations are significantly less than the government subsidies costs. This gives governments and/or industry the opportunity to augment and/or replace subsidies with this alternative model to encourage the uptake of EVs, make EVs more approachable, and reduce pollution in short journeys.

From a financial modeling perspective the most pressing problems, if indirect, relate to managing an implementation at scale, working within and refocusing current government and industrial policy, conveying the benefits and driving usage uptake with consumers, and technological advances that over time increase EV range while maintaining or decreasing EV pricing for the consumer.

10.5.4   Long-Term Simulation

We now present a simulation to further validate the performance of our car sharing system. We simulated demand based on the estimate for the probability of requiring an ICE vehicle from Figure 10.2 and serviced this demand using our car sharing and queueing model. Users requested an ICE (no accom-

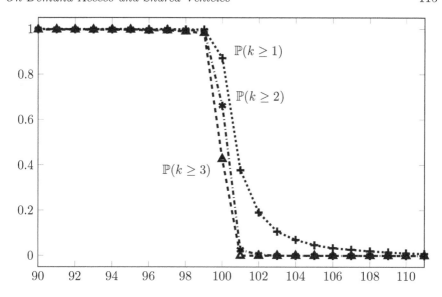

FIGURE 10.8
A simulation to illustrate the behavior of the queueing model. For bookings greater than three days in advance, the number of unhappy customers goes to zero.

modation for type of ICE was implemented in our simulation) with a fixed probability of $p = 0.1$ (to emulate infrequent need to take a long journey), independently of each other. The number of members of the car sharing program was $N = 1000$. The size of the ICE fleet was varied from $M = 90$ to $M = 111$. Figure 10.8 depicts the percentage of customers who experienced not being able to access a car within d_w days over a 1000 day period. Of course, a fleet of less than 100 ICE vehicles provided less than the average demand and did not result in a satisfying solution. The three graphs show (estimated) probabilities for a customer to have to wait (i) at least 3 days (dashed curve), (ii) at least 2 days (dashed-dotted curve) and (iii) at least a day (dotted curve). As can be seen, the number of unhappy customers (rejections) goes rapidly to zero as M exceeds 100 just by a small amount. In fact, for $M \geq 105$ no customers having to wait two days were observed and for $M = 111$, the chance of being rejected on the first day was below 1%. This is entirely consistent with the conclusions presented above and illustrates that the predictions of the mathematics are upper bounds on performance. We note that Figure 10.8 is a correction of the corresponding figure displayed in [106], where the figure shown does not correspond to the discussed values of N and p.

10.6 Reduction of Fleet Emissions

The sharing model described so far was applied to address the issue of customer range anxiety for electric vehicles. Now, we illustrate through a case study how the same idea can be applied to reduce fleet pollution levels. This objective becomes relevant when owners of fleets have a choice as to how EVs and non-EVs are needed to satisfy the needs of a business. By selecting the proportions correctly, one can reduce total fleet pollutants and guarantee a good quality of service for fleet users by ensuring flexible vehicle access based on statistical guarantees.

Large fleet owners (municipal authorities, universities, delivery companies) usually own various kinds of ICEVs for multi-functional purposes [99]. For example, in the case of a university, some of the vehicles are only adopted as campus vehicles with intermittent use, and others have longer travel demands based on continuous access. On any given day, these vehicles can be used for different purposes, for instance, security, cleaning, gardening and building maintenance. Some of these vehicles have fixed routines on campus (e.g., cleaning, postal), while other types of vehicles may have unpredicted driving requests (e.g., security vehicles for emergencies). Clearly, given the different demands placed on vehicles, there is a choice in the way we dimension this fleet. Electric vehicles are cleaner, and quieter, but ICEVs offer more flexibility in terms of driving range. The basic idea now is to replace the entire existing vehicle fleet with a mix of electric vehicles and ICE vehicles in order to reduce fleet emissions. We shall embed this idea in a statistical framework following the approach in the beginning of this chapter and use the solution presented there to dimension the number of ICE-based vehicles. Implicit in this assumption is again that the use of such vehicles can be planned a number of days in advance, and that all EVs can be fully charged overnight.

Then, based on the solution of the optimal mix in the shared fleet, we then briefly estimate the reduction of pollutants when compared with a fleet of only ICEVs. We do this by considering a specific case study as follows.

10.6.1 Case Study

We now consider the following scenario: a large company owns a large fleet of 200 ICEVs. This company wishes to update all of these vehicles to reduce pollutants. In the meantime, they also wish to know how many extra large sized ICEVs they need to buy for effective fleet management.

Here we adapt a real distribution of the driving distance range for the US postal services[6]. The profile for the distribution is illustrated in Figure 10.9. To be consistent with the assumption made in the postal report, it is con-

---

[6]https://postalmuseum.si.edu/research/pdfs/DA-WP-09-001.pdf. Last Accessed July 2017.

sidered that any delivery vehicle traveling more than 40 mile/d (64.37 km/d) is regarded as making a long journey, thus requiring large size ICEVs. From Figure 10.9, the probability for each user requesting an ICEV is calculated as 3.28%. We assume that the pollutant for each long-journey ICEV is taken as 350 g/km (see [144]). Then, the optimal values of M, when ϵ is 0.05 and for two different values of k are shown in Table 10.6. In Table 10.6 it is also possible to see that significant amounts of pollutants have been reduced, while guaranteeing a high QoS, even with a small proportion of ICEVs.

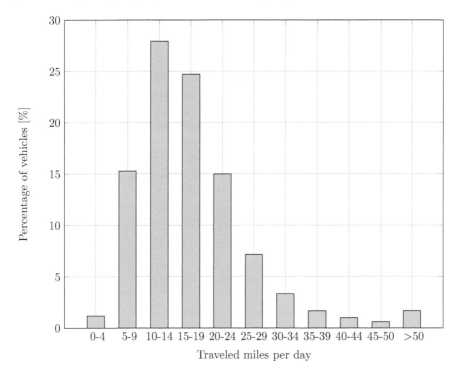

FIGURE 10.9
Distribution profile of the driving range for the vehicles in the delivery company

TABLE 10.6
Comparison Table for the Simulation Results

Total	k	ε	M	Reduced CO_2 (kg/day) [144]	Percentage CO_2 Reduced (%)
200	1	0.05	12	1594	81.87
200	3	0.05	8	1711	87.91

10.7   Concluding Remarks

In this chapter we have seen that there are various potential approaches to addressing the shortcomings of EVs as perceived by potential customers. In recent years there has been a significant drive to improve the range of EVs by increasing battery capacity, introducing more efficient production techniques for car components and motors. It may thus be argued that now or at least very soon an EV will easily provide the means of driving the standard routes of a large majority of car owners. However, some capabilities of ICEVs are clearly out of reach for EVs. The flexibility provided by an ICEV may therefore still outweigh the potential benefits of an EV.

A possible approach to this question is that OEMs see themselves not as car manufacturers but as providers of mobility concepts and means. The sharing-economy inspired concept proposed in this chapter would be to sell a mobility package which largely consists of an EV but which offers the capabilities of ICEVs as well for these relatively rare occasions where they are needed.

The overhead for such a solution is relatively small. Indeed, we have seen that significant amounts of pollutants have been reduced and good QoS can be provided with only a small proportion of ICEVs.

11

Sharing Electric Charge Points and Parking Spaces

11.1 Introduction

The previous chapter has considered designing shared services to alleviate the problem of range anxiety. An equally important problem is that of *charge point anxiety*; that is, the angst associated with the inability to access a charge point when charging is required. This is an important and active area of research for many innovative companies as well, see for instance the SmartPlug and the business model of BlockCharge[1], or the PlugShare app[2]. Charge-point access, and access to parking spaces, have much in common. Parking spaces represent a limited resource, for which competition is intense. Furthermore, the competition for spaces is often a result of artificial congestion, and this competition can sometimes be alleviated by sharing private spaces that are located close to areas where parking spaces are limited. As an example of such a scenario consider university campuses which experience parking congestion during the day, exactly at the same time when surrounding districts often have vacant spaces. In such situations home owners may be able to monetize their spaces by making them available to the university. Similarly, significant opportunities exist for owners of private charge points to monetize their charge points. The situation in the context of charge points is particularly frustrating in Ireland. Current policy (July 2016) is that purchasers of plug-in (the first 2000) EVs have a home charge point installed free-of-charge as part of an incentive scheme to encourage adoption of plug-in EVs. Thus, even though the number of available charge points in Ireland currently far exceeds the number of public charge points, remarkably, lack of infrastructure is still cited as an impediment to adoption of plug-in EVs. It is also worth noting that in the context of parking, major companies and cities are responding to these challenges. For example, SFpark[3] and JustPark[4] provide examples of companies

---

[1]https://www.linkedin.com/pulse/blockcharge-blockchain-based-solution-charging-cars. Last Accessed July 2017

[2]http://www.plugincars.com/how-to-use-plugshare-guide.html. Last Accessed July 2017

[3]http://sfpark.org. Last Accessed July 2017.

[4]https://www.justpark.com. Last Accessed July 2017.

investing in shared parking research and products within a smart cities context. Similar ideas are also currently emerging to share charge point access; see the initiative by Renault announced in 2016[5].

Our objective now is to discuss analytics to enable the design of such sharing systems. Specifically, we will describe the design of a system for sharing a resource so that spatial and temporal demand are better balanced. Recognizing that many private parking spaces and charge points are unused most of the day, and that resource congestion, or synchronized demand, is often caused by poor planning, we consider a situation where there are two nearby entities that have complementary supply and demand, i.e., there is a shortfall of resource at one and a simultaneous surplus at the other. In our work, we use the university campus as an example. The second entity is comprised of the homes of private residents who are in possession of their own garages, driveways and charge points, and these are made available as an overflow for the campus. Consequently, we consider a problem in which the campus has access to two classes of resource: a *premium* resource (parking spaces or charge point), e.g., those located on the university campus; and *secondary* resources located nearby and perhaps serviced by a shuttle. Given this basic scheme, we consider three specific design issues. First, we wish to design hardware to enable private home owners to allow other vehicles to use their driveway for charging or parking during pre-specified times. Second, we wish to guarantee a quality of service for the landlords by setting aside reserve resources on campus as contingency for events where secondary resources are suddenly unavailable. Finally, we wish to ensure that the premium resources are allocated optimally among *users* (drivers) while preserving each user's privacy. To keep matters simple, in what follows we shall illustrate this idea using parking spaces as the resource to be shared. We do this as data on the use of parking spaces is more readily available, and we have used this to develop a campus sharing systems that is documented in [76]. However, we note that the problem of charge points access can be solved in an identical manner, and our analytics apply equally well in such situations. Finally, the last part of this chapter describes a hardware device to enable the sharing of personal and private charge points.

11.2   Setting: Parking Spaces

Imagine a university campus with a total of N parking spaces, surrounded by a total of M private parking spaces (e.g., from residential complexes), as illustrated in Figure 11.1. A typical working day sees the university parking

---

[5]http://www.autovolt-magazine.com/swedish-citizens-create-electric-car-charging-infrastructure/. Last Accessed July 2017.

fill to capacity with vehicles belonging to students and staff to the extent that the N spaces cannot meet the demand. Some campus arrivals thus have to search elsewhere. At the same time, many nearby residents drive to work during the day and vacate their M private parking spaces. What we have is a wasted resource in one area (i.e. the residential parking spaces) and a stressed resource nearby (i.e., the on-campus parking).

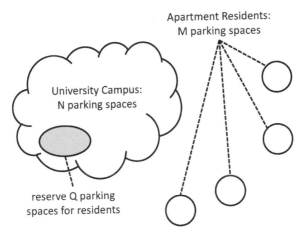

FIGURE 11.1
The parking scenario: premium spaces are those on the university campus, whereas secondary spaces are those belonging to the apartment buildings

We assume that a contract exists between the university and the landlords or owners of these M private parking spaces. The contract stipulates that these landlords lease their driveways to the university during the daytime while the landlords expect to be away (e.g., at work). In the sequel, we will denote the time period in which the extra spaces are leased by $[0, W]$.

We define premium resources to be the on-campus parking spaces, while the secondary resources are the contracted residential parking spaces. We will now consider how to design a parking system, bearing in mind the needs of the campus, and the needs of the landlords. We shall consider the following.

1. How do we accommodate the needs of parking space owners (landlords) given that situations may arise when landlords will need to return home during the contracted interval?

2. How do we accommodate the needs of the landlords given that situations arise when misbehaving university members will not vacate the parking space outside of the contracted interval?

3. How do we allocate efficiently access to the premium spaces to the university community in a manner that preserves the privacy of individuals in the community?

As we shall see, we answer these questions in a stochastic framework by developing suitable QoS metrics to dimension the aforementioned parking system. To address items 1 and 2, we shall set aside a subset of Q premium spaces as a reserve. We shall address item 3 by applying recent ideas from distributed resource allocation.

Remark 11.1 (Driveway Assumption). We assume throughout that the residential parking is such that each landlord has his or her own driveway as opposed to downtown parking scenarios with on-street parking only.

Remark 11.2. Note that the notions of *parking space* and *charge point* are interchangeable in the context of the above items.

11.3   Dimensioning and Statistics

In this section, we consider the problem of dimensioning $Q \in \{0, 1, \ldots, N\}$ premium spaces as reserve in order to provide sufficient QoS guarantees to the landlords of the secondary spaces. Let $1, \ldots, M$ index the parking spaces and suppose, for simplicity, that we consider landlords with a single driveway only. For each such parking space $i = 1, \ldots, M$, we define a non-negative random variable T_i which denotes the time at which he or she returns home and needs to get the parking space back. Under normal circumstances, T_i is greater than W, but on rare occasions a landlord may choose to come home early.

For simplicity, we assume that each parking space $i = 1, \ldots, M$ has exactly one daytime user per day. For each parking space $i = 1, \ldots, M$, we also define a non-negative random variable A_i which denotes the departure time of the daytime user of the space i. Under normal circumstances, A_i is less than W, but we assume a small number of miscreants so that not all spaces are always vacated on time.

For convenience, recall that $[0, W]$ denotes the nominal rental window for every parking space. In other words, the landlord of space i agrees to park only outside the interval $[0, W]$, whereas the daytime user of space i agrees to park during the interval $[0, W]$ only.

Definition 11.3 (Home-early and Overstay). We define the following two events for each secondary parking space $i = 1, \ldots, M$:

$$E_i \triangleq \{T_i \in [0, W]\} \cap \{T_i < A_i\},$$
$$O_i \triangleq \{W < T_i\} \cap \{T_i < A_i\}.$$

Each of these events represents an outcome where a landlord would like to use the space i but cannot do so. The *home-early* event E_i is due to the landlord needing the space during the day. The *overstay* event O_i is due to the fact that the daytime user overstayed.

For ease of presentation, we assume the following properties concerning the above random variables.

Assumption 11.4 (For Simplicity). The random variables $\{T_1, \ldots, T_M\}$ are independent and identically distributed (i.i.d). Likewise, $\{A_1, \ldots, A_M\}$ are i.i.d. Moreover, all these random variables are mutually independent. Finally, we assume that all the distributions have densities.

11.3.1 The Dimensioning Formulae

We begin by quantifying the probability of an event O_i in terms of the probability of daytime users overstaying. Recall that the distribution of the random variable A_i characterizes the probability of the daytime user overstaying in space i. First, observe that

$$\mathbb{P}(O_i) \leq \mathbb{P}(A_i > W).$$

Next, we derive an exact expression for $\mathbb{P}(O_i)$ using the independence assumption (Assumption 11.4).

Lemma 11.5 (Probability of O_1). *Let F_A denote the probability distribution of A_1 and let F_T denote the probability distribution of T_1. Under Assumption 11.4, we have*

$$\mathbb{P}(O_1) = \int_{a=W}^{\infty} (F_T(a) - F_T(W)) dF_A(a).$$

Remark 11.6. We estimate F_A from data in the following section.

Proof. Observe that

$$\mathbb{P}(O_1) = \mathbb{P}(W < T_1 < A_1)$$
$$= \int_{a=W}^{\infty} \mathbb{P}(W < T_1 < A_1 \mid A_1 = a) \, dF_A(a)$$
$$= \int_{a=W}^{\infty} \mathbb{P}(W < T_1 < a) \, dF_A(a).$$

The claim follows by definition of the distribution F_T of T_1. □

Now we derive a formula that we can use to dimension the reserve parking space Q from the contingent of N premium spaces. Recall that there are Q reserve parking spaces (the reserve "buffer") set aside by the university. We consider the probability $p(M, Q)$ of the event that more than Q spaces are needed to accommodate landlords needing the reserve buffer during daytime $[0, W]$:

$$p(M, Q) = \mathbb{P}\left(\sum_{i=1}^{M} 1_{E_i} > Q\right),$$

where 1_{E_i} denotes a Bernoulli random variable taking the value 1 when event E_i occurs, and the value 0 otherwise. We first characterize the probability of the event E_1 in terms of the probability distributions of T_1 and A_1.

Lemma 11.7 (Probability of E_1). *Let F_T denote the probability distribution of T_1. Let F_A denote the probability distribution of A_1. Under Assumption 11.4, we have*

$$\mathbb{P}(E_1) = \int_{t=0}^{W} (F_A(W) - F_A(t))dF_T(t) + F_T(W)(1 - F_A(W)).$$

Remark 11.8. F_T and F_A are estimated from data in the next section.

Proof. Let $\phi \triangleq \mathbb{P}(E_1)$. Observe that

$$\begin{aligned}
\phi &= \mathbb{P}(\{T_1 < A_1\} \cap \{T_1 \in [0, W]\}) \\
&= \mathbb{P}(\{T_1 < A_1\} \cap \{T_1 < W\}) \text{ (by non-negativity of } T_1) \\
&= \mathbb{P}(\{T_1 < A_1\} \cap \{T_1 < W\} \mid A_1 \leq W)\mathbb{P}(A_1 \leq W) \\
&\quad + \mathbb{P}(\{T_1 < A_1\} \cap \{T_1 < W\} \mid A_1 > W)\mathbb{P}(A_1 > W) \\
&= \mathbb{P}(T_1 < A_1 \mid A_1 \leq W)\mathbb{P}(A_1 \leq W) \\
&\quad + \mathbb{P}(T_1 < W)\mathbb{P}(A_1 > W) \text{ (by Bayes' Rule)} \\
&= \mathbb{P}(T_1 < A_1 \leq W) + \mathbb{P}(T_1 < W)\mathbb{P}(A_1 > W) \\
&= \int_{t=0}^{W} \mathbb{P}(t < A_1 \leq W)\,dF_T(t) + \mathbb{P}(T_1 < W)\mathbb{P}\mathbb{P}(A_1 > W) \\
&= \int_{t=0}^{W} (F_A(W) - F_A(t))dF_T(t) + F_T(W)(1 - F_A(W)),
\end{aligned}$$

which is the claim. $\square$

As a corollary, we have the following expression for the probability that setting Q reserve spaces at the university is not enough.

Corollary 11.9 (Probability that Q reserve spaces are not enough). *Let $\phi \triangleq \mathbb{P}(E_1)$. Under Assumption 11.4, $p(M, Q)$ is a random variable entirely characterized by ϕ:*

$$p(M, Q) = \sum_{k=Q}^{M} \binom{M}{k} \phi^k (1 - \phi)^{M-k}. \tag{11.1}$$

11.3.2 Parking Data and Example

In the previous section, we considered arbitrary probability distributions F_T and F_A in the formulae derived. In this section, we give estimates $\hat{F}_T, \hat{F}_A$ of these distributions using publicly available data. Given samples $Z_1, Z_2, \ldots, Z_n$ from the distribution F_T, the corresponding empirical distribution-estimate takes the form

$$\hat{F}_T(z) = \sum_{i=1}^{s} 1_{[Z_i \leq z]},$$

where s is the sample size. First, we estimate the distribution F_A. Recall that A_i is the random variable denoting the duration of use of the ith secondary parking space. To derive an estimate, we use data on parking space utilization collected in the city of Dublin. Each data point corresponds to the time duration of one parking event. The histogram distribution of these durations is shown in Figure 11.2. Of course, we can obtain a better estimate of the distribution of parking usage in a university campus if we have access to more particular data. Based on Figure 11.2, in order to simulate the fact that 5% of users of secondary parking spaces overstay, we set $W = 170$ for our example.

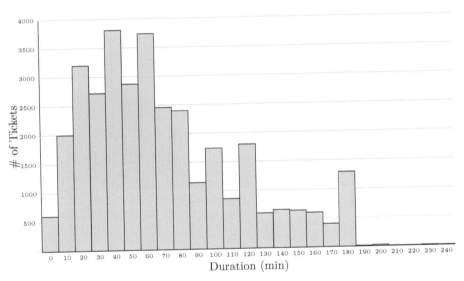

FIGURE 11.2
Parking data (Source: Dublin City Council, 21 December 2013)

Next, we estimate the distribution F_T for $\{T_i\}$. This distribution accounts for landlords who do not leave home and who arrive home normally after working hours. There are many reasons for landlords to return home early, or not leave home at all. For simplicity, we estimate the frequency of such days with the number of sick days had by NHS staff in England over the period from April 2009 to February 2014. This data is presented in Figure 11.3. For simplicity, we use the average sickness absence rate to estimate the probability of $T_i = 0$ and assume that $T_i = W$ otherwise (see Table 11.1).

If data is not available to estimate the distributions F_T and F_A, it is

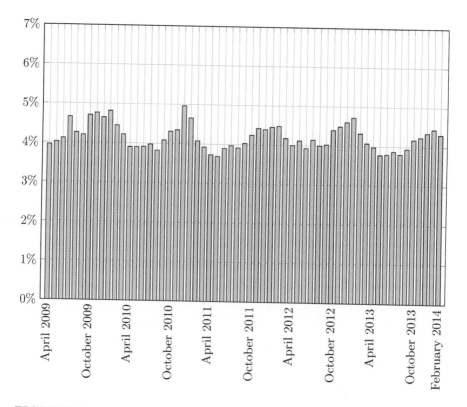

FIGURE 11.3

Monthly sickness absence rates (Source: Health and Social Care Information Centre)

TABLE 11.1

Probability distribution function of T_i. For simplicity, we assume that the function is constructed so that 4.2% of landlords remain at home due to sickness, while the other 95.8% return home exactly at time instant W.

Random variable T_i	Probability
$T_i = 0$	0.042
$T_i = W$	0.958

insightful to consider the assumption that F_T and F_A are normal distributions $N(\mu, \sigma^2)$ and $N(\nu, \rho^2)$, respectively. This assumption gives a reasonable model for F_T and F_A as long as the tail probability in the negative half line is negligible, e.g. when the variances are small compared to the means. In this case, letting Φ denote the standard normal distribution function, Lemma 11.7 gives:

$$\mathbb{P}(E_1) = \int_0^W \left(\Phi\left(\frac{W-\nu}{\rho}\right) - \Phi\left(\frac{t-\nu}{\rho}\right) \right) e^{-\frac{(t-\mu)^2}{2\sigma^2}} dt$$
$$+ \Phi\left(\frac{W-\mu}{\sigma}\right) \left(1 - \Phi\left(\frac{W-\nu}{\rho}\right) \right).$$

By further assuming that the expected return time μ of landlords is 2ϵ above the expected departure time ν of daytime users, and that the rental term W is the midpoint between μ and ν, we obtain

$$\mathbb{P}(E_1) = \int_0^{\nu+\epsilon} \left(\Phi\left(\frac{\epsilon}{\rho}\right) - \Phi\left(\frac{t-\nu}{\rho}\right) \right) e^{-\frac{(t-\mu)^2}{2\sigma^2}} dt$$
$$+ \Phi\left(\frac{-\epsilon}{\sigma}\right) \left(1 - \Phi\left(\frac{\epsilon}{\rho}\right) \right).$$

We can therefore conclude that for sufficiently large ν compared to the variances σ^2 and ρ^2, the value of $\mathbb{P}(E_1)$ depends mainly on the value of ϵ, or the difference between μ and ν.

Finally, we use Corollary 11.9 to perform a dimensioning exercise based on our data. Figure 11.4 illustrates the probability $p(M, Q)$ that Q reserve spaces are insufficient when M secondary spaces are contracted. For a fixed value of $\mathbb{P}(E_1)$, the probability $p(M, Q)$ eventually falls exponentially fast versus Q. The dependence of $p(M, Q)$ on M is more subdued. In other words, for a fixed value of $p(M, Q)$, a linear increase in the number of secondary spaces only requires a logarithmic increase in the number of reserve spaces.

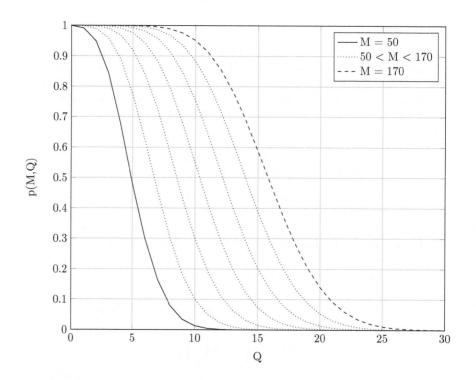

FIGURE 11.4
Probability $p(M, Q)$ that Q reserve spaces are insufficient when M secondary spaces are contracted

11.4 Efficient Allocation of Premium Spaces

In this section, we focus our attention on the point of view of the user (i.e. a staff member or student) who requires parking at the university. Typically, such users would purchase a monthly or yearly parking ticket. This ticket then provides them with the opportunity to compete for a university parking space. However, such a ticket does not necessarily guarantee them a parking space on the university grounds. That is, if they arrive to the university "too late" on any given day, the car parks on campus may already be full. This is because more parking tickets are sold than there are actual spaces to park. Such "first come, first served" systems can be inefficient. Consider the example of student mothers or fathers who, even though they paid the same amount for a monthly or yearly parking ticket as everyone else, must first always take their children to school in the morning.

For our scheme, we suppose that the university has been able to obtain enough apartment building parking spaces such that $(N - Q) + M$ is greater than or equal to the number of university users requiring a parking space on any given day. The problem then is how to allocate the premium and secondary parking spaces to users efficiently and fairly over time.

11.4.1 Algorithm

We are considering the problem of providing efficient access to the available premium spaces on campus. These spaces refer to the $N - Q$ spaces available after allocating Q of the spaces as reserve according to Section 11.3. For simplicity of exposition, we shall assume that $Q = 0$ in this section, but the results generalize in a straightforward fashion.

We introduce and recall the following notation to describe the problem data and the variables used in our algorithm.

N : an integer denoting the number of available premium parking spaces.

n : an integer denoting the number of users wishing to use premium parking spaces. We assume that $n > N$.

k : denotes discrete time, $k = 0, 1, 2, 3, \ldots$. In our interpretation, this corresponds to the number of days the system is operating. For convenience, we assume spaces are assigned on a per-day basis, but the general principles of the algorithm do not depend on this assumption.

$X_i(k)$: this is a state variable associated with the ith user. It takes the value 1 if this user is given access to a premium parking space on the kth day and zero otherwise.

$\overline{X}_i(k)$: this is average access for the ith user up to the kth day, i.e.

$$\overline{X}_i(k) = \frac{1}{k+1} \sum_{j=0}^{k} X_i(j).$$

It is possible to formulate the premium parking space allocation problem in several ways. For example, one could require that the long-term average admission to the premium parking space is equal for all users, i.e., for all $i, j = 1, \ldots, n$,

$$\lim_{k \to \infty} \overline{X}_i(k) - \overline{X}_j(k) = 0.$$

This assumes that all users are equal in the desire to access the premium parking space. Here, we follow a more general approach and assume that each user i has a cost function $f_i : [0, 1] \to \mathbb{R}$. For a frequency $z \in [0, 1]$ of premium space allocation, the cost $f_i(z)$ represents the monetary inconvenience cost that user i experiences from z. This function specifies the priority that this user is assigned. It could represent the amount a user is willing to pay, or it could be related to the number of passengers carried by this user, or the access that this user has to public transportation (meaning that users with fewer possibilities for alternative transport should have prioritized access to parking spaces). Given these individual cost functions, our aim is to design a system that achieves overall minimal cost for the group. We formulate the optimal allocation of resources as a minimization problem:

$$
\begin{cases}
\displaystyle\min_{z_1,\ldots,z_n \in \mathbb{R}} & \displaystyle\sum_{i=1}^{n} f_i(z_i) \\
\\
\text{subject to} & \displaystyle\sum_{i=1}^{n} z_i = N, \quad \text{and} \quad z_i \geq 0, \quad i = 1, \ldots, n
\end{cases}
\tag{11.2}
$$

Our proposed simple algorithm for solving the parking allocation problem can be summarized as follows. We assume that, for each day, each user is allocated access to the premium spaces by tossing a coin. For example, one embodiment of this idea is to use a smart-phone application. More specifically, each user is assigned a cost function by a government authority. For example, this could be based on vehicle class, disability, need for childcare, etc.

We will assume that the functions f_i are continuously differentiable and strictly convex so that, in particular, the optimal point $z^* \in \mathbb{R}^n$ satisfying the constraints is unique. Furthermore, we introduce an assumption which ensures that the optimal point z^* has only positive entries. This assumption also guarantees that the algorithm we will describe is well defined for every user.

We wish to control access to the premium space in such a way that the average utilization for each user approaches the optimal value z^*, i.e. for large k we want to achieve

$$\overline{X}_i(k) \approx z_i^*$$

subject to the (loose) capacity constraint $\sum_{i=1}^{n} X_i \approx N$. That is, all premium spaces are occupied on average. Furthermore, we wish to do this in a manner that preserves the privacy of users. That is, we do not wish to reveal $\overline{X}_i$ and f_i to any other user during the course of the optimization. Finally, the necessary communication between the users should be minimal so as not to create a communication overhead that would be hard to sustain in an uncertain environment where users cannot be expected to participate at all times.

In what follows, the mechanism for preserving privacy is to develop a distributed algorithm. We loosely follow the ideas in [192], where a distributed stochastic algorithm is presented which guarantees that the average utilization variables $\overline{X}_i(k)$ converge to the optimal points z_i^{*}[6]. Also, a more detailed discussion on the distributed optimization algorithm AIMD, and its convenience for large scale applications, is given in Chapter 18. The algorithm presented here extends the ideas of [192], however, as we need to address the further constraint that the instantaneous utilization variables $X_i(k)$ sum to N, or at least to a value close to it. Moreover, the resource to be allocated in our setting is atomic as opposed to arbitrarily divisible. These differences require substantial changes to the algorithm presented in [192].

At each time k, each user i determines a probability $p_i(k)$ and sets

$$X_i(k+1) = \begin{cases} 1, & \text{with probability } p_i(k), \\ 0, & \text{with probability } 1 - p_i(k), \end{cases} \tag{11.3}$$

where we note that all users make this probabilistic choice independently of other users or previous decisions. The evolution of the probabilities is governed by the equation

$$p_i(k) \triangleq \mathbb{P}(X_i(k) = 1) = \Gamma(k) \frac{\overline{X}_i(k)}{f_i'(\overline{X}_i(k))}. \tag{11.4}$$

Note that each user i can determine its own probability with the exclusive knowledge of its own past utilization $\overline{X}_i(k)$ and cost function f_i. No information from other users is required. The scalar $\Gamma(k)$ is a network-wide constant determined by the central agency and broadcast to all users. Here, $\Gamma(k)$ is chosen such that $p_i(k) \in (0,1)$ for all $i = 1, \ldots, n$ and all $k \in \mathbb{N}$; for instance, we may assume that the central authority that owns the parking lot calculates $\Gamma(k)$ based on past utilization and broadcasts this scalar to all participating vehicles. It is determined in a time-varying manner as it also influences the demand for premium spaces. Specifically, if at a certain time k each $p_i(k)$ is fixed then the expected utilization of the premium spaces is

$$\mathbb{E}\left(\sum_{i=1}^{n} X_i(k+1)\right) = \sum_{i=1}^{n} p_i(k) = \Gamma(k) \sum_{i=1}^{n} \frac{\overline{X}_i(k)}{f_i'(\overline{X}_i(k))}. \tag{11.5}$$

---

[6] As the algorithm is stochastic, the convergence holds with probability 1, which is also called almost sure convergence in a stochastic context.

Moreover, the (random) instantaneous utilization $\sum_{i=1}^{n} X_i(k+1)$ is concentrated around the expected utilization by independence and Hoeffding's Inequality [93]. If we wish to ensure optimal utilization of the premium spaces and avoid overbooking, the expected utilization should be below the number of premium parking spaces; for instance, a standard deviation below this number, depending on the desired QoS metric. Denoting this number by $N_E \leq N$, we will therefore adjust $\Gamma(k)$ so that the expectation in (11.5) tracks N_E. As the expectation is unknown, we use the observed utilization as an estimator for this. Taking a simple error regulation approach, we thus arrive at

$$\Gamma(k+1) = \Gamma(k) + \gamma \left(N_E - \sum_{i=1}^{n} X_i(k) \right). \tag{11.6}$$

The overall system is now described by the dynamics of X_i given by (11.3), the dynamics of p_i given by (11.4) and the dynamics of Γ as in (11.6).

11.4.2 Example

We simulate a population of 900 users competing for 450 premium parking spaces. Each evening, users are assigned a parking space as described above (with the scalar $\Gamma(k)$ determined using the simple error regulation law (11.6). For simplicity, users have one of three cost functions: $f_1(z) = 1 - z + z^4/4$, $f_2(z) = 1 - z + z^6/6$, and $f_3(z) = 1 - z + z^8/8$, all of which are strictly convex. Also the highest cost is associated with $z = 0$ which fits the interpretation of cost incurred for not obtaining access to a premium parking space. For better interpretation of the constant $\Gamma(k)$ we simulated the problem for the normalized cost functions $\tilde{f}_i(z) = f_i(z) - 1 + z, i = 1, 2, 3$. This does not change the optimal point by the Karush-Kuhn-Tucker (KKT) conditions. However, it ensures that $\Gamma(k)$ is positive.

Figure 11.5 shows the average utilization achieved for each class of vehicle. Figures 11.6 and 11.7 show that the average utilization of premium spaces is concentrated around the target utilization of 450.

11.5 Turning Private Charge Points into Public Ones

The previous discussion exploited the strong parallels that exist between the *angst* associated with not being able to find a parking space in a congested environment, and the (perhaps) extreme angst that EV owners feel when attempting to charge their vehicles in a non-home environment. The solution advocated in the prequel to both of these problems is to allow owners of parking spaces or charge points to make these available to others in exchange for monetary reward. Note again that the idea of sharing charge points is well founded

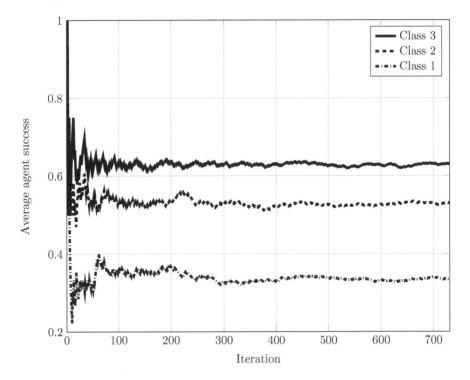

FIGURE 11.5
Instantaneous allocation $\{X_i(k)\}$ for three users

for a number of reasons. First, in Ireland, the geographical density of public or community charge points is often very low leading to congested behavior in certain areas. This is especially true in campus environments (universities, business parks, large scale industrial complexes). In addition, since charging times are much longer than the process of refueling petrol, long queues may build up at some charging stations, thus further reducing the practical availability of charging points. Thus, demand for public charge points often outstrips supply and in such situations it makes sense to augment the public charge points with private ones. Second, in Ireland, public charge points are currently deployed by the State (via the ESB Group), and accessed for free by registered EV owners who qualified for a state grant to purchase their electric vehicle. It is planned that this free access will end at some point, after which users will pay to charge their vehicles. Furthermore, in Ireland, purchasers of plug-in vehicles have had a home charging unit installed at their homes as part of the EV incentive scheme. Thus, in addition to the public charge points, a network of private charge points is not only available to augment, but also to compete with the public network and to enable a market for EV charging in Ireland. Note also that while such sharing strategies are very easy

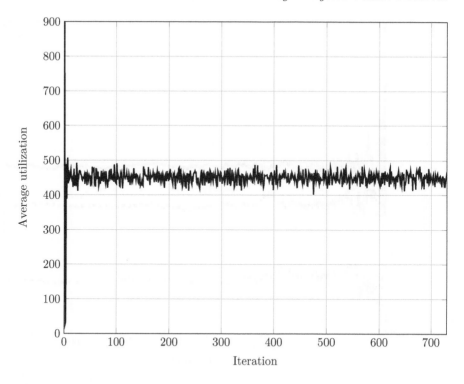

FIGURE 11.6
Instantaneous premium space occupancy over time

to implement in the case of parking, requiring very little in the way of new infrastructure, unfortunately this is not the case for sharing of charge points. A typical 16 A home charging unit (available as part of the ESB scheme) is depicted in Section 11.5.

As can be seen, access to the unit is managed via a key and the car simply plugs into the device until charged. To enable smart sharing, we have constructed a smart adapter[7]. The adaptor is designed to convert a standard IEC 62196 16 A home charging unit into a public one. A schematic for the device is depicted in Figure 11.9. As can be seen the smart plug acts as an interface between the plug-in EV and the home charging station. Charging of multiple vehicles is supported (via multiplexing); and the device connects to the cloud via the cellular network or WiFi. The amount of charge delivered to a vehicle is monitored internally, and several payment modes are supported; direct payment; a subscription model; or via a web interface. The charging cable at the car remains locked until payment or complete charging. More details of the unit can be also found in [181].

---

[7]*Pricing adapter for domestic EV charging stations*, United Kingdom Patent Application No. 1621894.3.

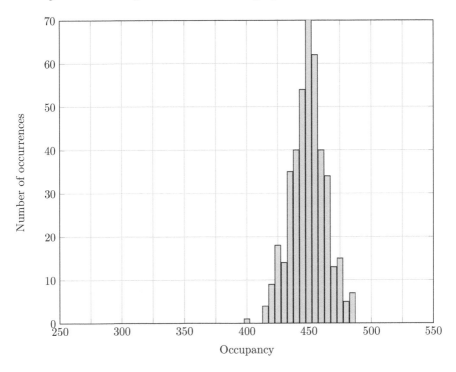

FIGURE 11.7
Histogram of premium space occupancy

11.6 Concluding Remarks

The lack of infrastructure is sometimes cited as an impediment to the adoption of plug-in EVs, as it may give rise to the fear of not finding a charge point when required. However, as we describe in this chapter, this fear may be alleviated by simply improving the efficiency of the existing infrastructure, for instance by designing shared strategies to access it. In particular, we have shown how a smart adapter may be to convert a standard home charging unit into a public one. Then, after recognizing that most private charge points are unused most of the day, when owners are at work, we discussed analytics to enable the sharing of the private charge point, with a probabilistically guaranteed quality of service.

FIGURE 11.8
Home charging unit in Ireland

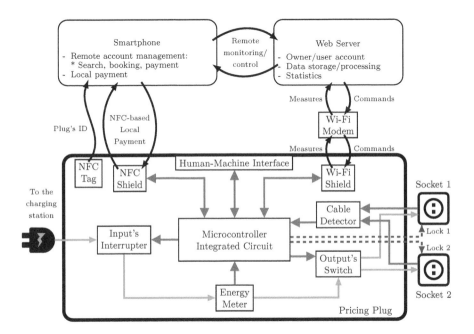

FIGURE 11.9
Smart charging unit

Part III

EVs and Smart Cities

12

Context-Awareness of EVs in Cities

12.1 Introduction

In this part we describe a number of non-conventional ancillary services that can be offered by EVs. Conventional ancillary services usually focus on the services that can be offered by plug-in EVs when they are connected for recharging, and they usually include demand response, frequency regulation, and reactive power balancing. Such possibilities have been explored in the first part of this book. On the other hand, in this part we are interested in exploring the services that can be provided by EVs when they are traveling.

The first chapter is devoted to the analysis of the services that can be delivered by hybrid vehicles in particular. In this case, the principal concern of the hybrid architecture is usually the optimization of the fuel efficiency of the vehicle. This implies that the Engine Management Unit (EMU) is programmed to minimize fuel efficiency, and maximize the expected lifetime of the vehicle battery. However, the hybrid architecture can be exploited to optimizing the performance of the vehicle with respect to other stakeholders (e.g., pedestrians, society in general). For instance, one can imagine a PHEV as a moving agent that has the ability to decide when, and where, to pollute (i.e., travel in an "ICE-like" mode, where most or all of the torque is produced by the combustion engine). On the other hand, the PHEV can turn into a "FEV-like" mode, when most or all of the torque is produced by the electric engine, when and where it is preferable. In this way, the PHEV can drive in an environmentally friendly fashion in sensitive areas, e.g., close to kindergartens or hospitals, or in the city center. In this regard, note that some cities are planning an outright ban on polluting vehicles from the city center.

In a similar manner, the ability of the EMU to orchestrate between two traveling modes can be exploited to offer even more services. In Chapter 14, PHEVs decide when and where to travel in a FEV-like mode in order to discharge their batteries proactively of a pre-fixed quantity. In this way, the PHEV becomes a deterministic load from the viewpoint of the power grid, and thus, a schedulable and dispatchable load. Accordingly, this mitigates the burden of the grid to accommodate an otherwise unpredictable and possibly non controllable load. In addition to this, it is possible to use the PHEVs to store inexpensive energy generated from renewable sources. Roughly speaking, if weather forecasts predict a large availability of energy from wind plants for the next night, when PHEVs will be connected for charging, then it is possible to suggest to the PHEV's owner to consume most of the energy stored in the battery during the day. Such an approach is also illustrated for the specific case of a fleet of hybrid buses.

Finally, Chapter 15 describes an Intelligent Speed Advisory (ISA) system for EVs. In general, ISA systems are used to provide a number of benefits, including an improved vehicle and pedestrian safety, a better utilization of the road network, or reduced emissions. Here, we are interested to the specific case of EVs, and provide a methodology that allows drivers to manage a budget of some description, and at the same time maximize the energy efficiency of their EVs when traveling in a green zone area.

Notes and references

This part is based on joint work by the authors and their co-workers. The chapter on the regulation of pollution is based on [164] written in collaboration with Arieh Schlote, Florian Häusler, Thomas Hecker, Astrid Bergmann and Ilja Radusch[1]. The chapter on the controlled energy consumption on the basis of weather forecasts is based on works [80] and [137], written in collaboration with Yingqi Gu, Florian Häusler, Wynita Griggs, Joe Naoum-Sawaya and Mingming Liu[2]. Finally, the chapter on the ISA system is based on works [122][3] and [123][4] in collaboration with Mingming Liu, Rodrigo Ordóñez-Hurtado and Yingqi Gu.

---

[1]©IEEE. Reprinted, with permission, from [164].
[2]©IEEE. Reprinted, with permission, from [137].
[3]©IEEE. Reprinted, with permission, from [122].
[4]©IEEE. Reprinted, with permission, from [123].

13

Using PHEVs to Regulate Aggregate Emissions (twinLIN)

Our objective in this chapter is to describe an attractive method to regulate pollution without overly inconveniencing a vehicle owner. The basic idea is to create a feedback loop that, based on measurements of pollution levels, gives recommendations to the EMUs of PHEVs regarding the driving mode, in order to maintain the pollution level below a safe pre-determined level. Traditionally, this can be achieved by adapting speed limits, re-routing vehicles, and by changing traffic light sequencing. These measures are highly invasive. However, PHEVs, offer this new flexibility in terms of the driving mode that can be exploited to solve the pollution regulation problem, potentially, without overly inconveniencing the vehicle owner. Such a vision corresponds to viewing the battery as a type of filter for vehicular traffic that geographically separates the location where energy is used and the location where it is being produced (perhaps using fossil fuels or other "dirty" forms of energy). Essentially we generate the energy in a place that is away from humans, so that pollutants can be filtered, gathered and neutralized, and we deliver it, via batteries, in a form that is clean and safe.

When taking this point of view, new vehicle classes such as electric and hybrid electric vehicles that allow cars to traverse sensitive areas without polluting them, become a powerful tool in controlling pollution levels in cities. Power-split hybrid vehicles, in particular, which can be operated in fully electric, and in ICE mode, allow us to control the manner in which pollution is delivered into the environment. Thus by orchestrating the way in which a fleet of such vehicles switch into fully electric mode (based on a function of the aggregate pollution levels), one should, in principle, be able to regulate pollution levels in a manner that is non-invasive to the driver. The vehicle only uses as much electric power as is necessary to keep the aggregate pollution level below a certain threshold. More refined versions of this basic idea can be used to specifically protect cyclists and to protect pedestrians [81, 91]. This chapter explains how this can be achieved in practice, by describing the construction of a proof-of-concept context aware hybrid vehicle: the *twinLIN*, and by describing the application of basic control theory to this problem, and is based on prior work in [164], [110] and [109] which investigated methods based on V2X (Vehicle-to-Everything, e.g., V2V and V2I) to contribute to the regulation of air quality in our cities.

Before proceeding it is worth noting that the link between the ICE and human health has recently become a very hot topic. Recent papers link the harmful by-products of the ICE to a range of ailments. By-products of the ICE include: CO; NOx; SO; $Ozone$; $Benzene$; $PM10$; and $PM25$. All of these affect humans adversely and have been linked to lung disease, heart disease, certain cancers, and most recently dementia [33] [66]. In a study in the US [38] it is claimed that problems with air quality lead to three times as many deaths as car accidents. Amazingly, just how damaging to health those vehicles can be appears only now to be a topic of interest with public discourse hitherto focusing mainly on greenhouse emissions and on vehicle safety. This basic fact provides the main motivation for this and prior work [110] and [109]; namely, to investigate methods based on V2X to contribute to the regulation of air quality in our cities.

It is also worth noting that governments and municipal authorities have already started to respond to the air-quality issue. For example, cities in some countries ban certain vehicles from densely populated areas (*Umweltzonen*[1]), and sometimes speed limits are adapted to respond to pollution peaks[2]. In particular, the concept of the *Umweltzonen* is widespread throughout Germany. Limits on particulate matter and other pollutants have been in effect in Germany and the EU for some time. For example, in the EU, exposure to a yearly average of $40\,\mu g/m^3$ and a daily average of $50\,\mu g/m^3$ have been set for particulate matter smaller than $10\,\mu m$ (PM10). However, even though these measures are very welcome, they do not go far enough. Roughly speaking, they suffer from three main drawbacks. First, they are per-vehicle measures. Degradation of air quality results from the aggregate effect of vehicles. Enforcing per-vehicle measures (unless we ban vehicles all together), takes no account of this effect. In fact, while the per-car emissions have been successfully decreased in the last years, the growth in the number of new cars has led to a substantial effective increase in the overall (aggregate) emission output in certain regions [54]. Second, these measures are open-loop measures. The regulation is the same irrespective if there is one vehicle in a spatial area (in the middle of the night), or if there are thousands of vehicles in the same area. Finally, all of these measures are highly invasive and affect the vehicle owners in a very disruptive manner. They may even have unintended consequences for the city. Forcing vehicles away from an certain zone may lead to congestion elsewhere or even a higher total amount of emissions in the whole city.

This chapter is organized as follows: In Section 13.1 we present some important notions from networking research and describe how these translate in a seamless fashion to fleets of hybrid vehicles, by using some V2V and V2X communication. In Section 13.2.1 we describe the construction of a proof-of-concept context aware hybrid vehicle, the *twinLIN*, and the smart-phone application we used to implement the previous idea. In Section 13.2.2, we

---

[1]http://gis.uba.de/website/umweltzonen/umweltzonen.php. Last Accessed July 2017.

[2]http://www.brussels.be/artdet.cfm/6357. Last Accessed July 2017.

review a pollution model (that we have used in our models in the absence of pollution measurements). We later present some simulation results that show the efficacy of the proposed idea, at least in a simulation environment, and outline some possible more sophisticated generalizations of the proposed framework.

13.1   Background

The principal objective of this chapter is to use the opportunity afforded by V2X to develop effective techniques to use PHEVs to regulate the pollution level in urban areas. The basic idea in [109, 164] is to place a feedback loop around a group of PHEVs, and use this loop to control the group emissions. Specifically, we are interested in the following aspects:

(i) Aggregate pollution levels should not exceed some given levels: More specifically, the objective here is to decouple pollution in a certain area (or group of vehicles) from the number of vehicles in that area.

(ii) Best effort behavior and disturbance rejection: Vehicles in a geographic area should adjust their behavior in order to share the allowed pollution level irrespective of the number of participating vehicles and should also respond to mitigate non-vehicle pollution generation by becoming cleaner if pollution levels rise.

(iii) Fairness : Ideally, cars that are more polluting should be more inconvenienced than less polluting ones.

Items (i)-(iii) closely resemble the requirements of classical resource allocation algorithms found in networking applications (such as the Internet). Essentially, we are viewing the pollution control problem as a resource allocation problem, where a certain amount of pollution is shared among competing vehicles, see for instance Figures 13.1 and 13.2. The realization that the pollution control problem can be recast in a resource allocation framework is very fortunate. Network resource allocation problems are at a mature stage, and algorithms from the networking community for solving such problems are readily available. Relevant ideas include the Kelly framework [101], RED [167], AIMD congestion control, to name but a few. In the next section we shall illustrate how some of these ideas, together with new vehicle types, can be used to great effect to manage aggregate vehicle emissions.

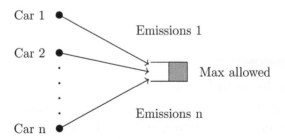

FIGURE 13.1
Pollution as a resource allocation problem

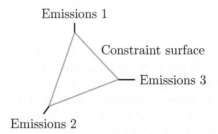

FIGURE 13.2
Pollution control involves controlling the levels of emissions of different pollutants at the same time

13.2 Cooperative Pollution Control

The basic idea now is to orchestrate and coordinate switching between drive modes in a fleet of hybrid vehicles so as to achieve regulated pollution levels. Feedback is used to adjust the level of coordination to achieve the desired level of regulation. The control loop is depicted in Figure 13.3. In our work, the infrastructure uses algorithms of the form of Algorithm 13.1, where $g(\cdot)$ is a function of present and past emissions and depends on the chosen control algorithm.

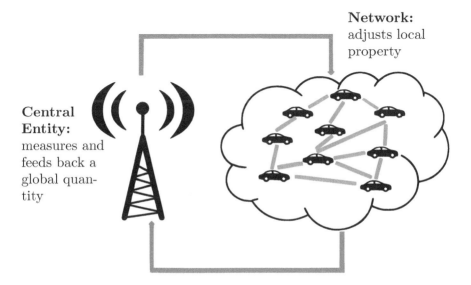

FIGURE 13.3
Feedback loop for cooperative pollution control.

Algorithm 13.1 Central probability control

$E(k) \leftarrow$ Measurement or estimate of current emission levels
$p \leftarrow g(E(k), E(k-1), ...)$

Broadcast probability of changing to EV mode (p) to all vehicles.

To realize this objective we assume the availability of a context-aware hybrid vehicle, whose switching into fully electric mode can be made dependent on the location of the vehicle as well as in response to an external signal. Our approach here is to allow vehicles to randomly select their mode of operation in response to these signals (in contrast to the approach in [110]) as in Algorithm 13.2. Here $\hat{p}$ is the probability that a hybrid electric vehicle engages its ICE engine and $f(\cdot)$ is a function used internally in the vehicle to allow one of several different types of optimality objectives to be achieved. Another important point that has to be considered is the availability of accurate and real time pollution estimates or measurements. These can be obtained in two ways: either by roadside infrastructure measurements or by communication of on-board measurements of emissions of the vehicles themselves.

Algorithm 13.2 Vehicle mode choice

$p \leftarrow$ broadcast probability value
$\hat{p} \leftarrow f(p)$

Use ICE with probability $\hat{p}$ and EV mode otherwise.

We now describe all the ingredients required to implement the desired cooperative pollution control task.

13.2.1 The Networked Car

The first vehicle that we used for cooperative pollution control was a 2008 model Toyota Prius, as described in [164]. More recently, a 2015 Toyota Prius VVTi 1.8 5DR CVT Plug-in Hybrid vehicle was used in its place. The engine management system of the Prius allows the vehicle to be powered by the ICE alone, the battery, or using a combination of both, and it is this degree of freedom that can be used to regulate emissions. For such an objective, we have made some important modifications to the basic vehicle to make it behave as a context-aware vehicle. First, we have automated the switching of the vehicle from ICE to EV mode by adapting the EV-mode button hardware in the vehicle. For this purpose, a dedicated Bluetooth-controlled mechanical interface was constructed to override the manual EV button based on signals from a smartphone. The switching can be triggered on the basis of the GPS location, of external context information, or on-board signals such as speed and battery level. Second, special-purpose hardware was constructed to permit communication between a smartphone and the controller area network (CAN) bus. The Prius provides a CAN access to the vehicle diagnosis (On Board Diagnosis II (OBDII)) interface. Using special commands on the Bluetooth channel, the smartphone subscribes to specific in-vehicle data. The CAN Gateway reads this data from the CAN interface, filters the requested data and delivers them to the smartphone. Due to safety reasons, the module works in a read-only mode as CAN networks are very sensitive to adaptations. Communication to other vehicles, to GPS, and to a cloud server can be also realized through the smartphone. Although different choices are obviously possible, in our latest tests we use a Samsung Galaxy S III mini (model no. GT-I8190N) running the Android Jelly Bean operating system (version 4.1.2) and the OBDII interface device that we used was the Kiwi Bluetooth OBD-II Adaptor by PLX

Devices[3]. Thus, the smartphone has access to in-vehicle bus data as driving mode, battery level, and pollution levels via a central server. The Android application then allows the vehicle to interact with its environment in a very smart manner. Importantly, it allows controlled delivery of emissions and pollution into the city environment (be it noise pollution, or more directly harmful pollutants), allowing us to control where and when these emissions are delivered into the environment. A movie demonstrating the first operations of the vehicle can be found at `http://www.hamilton.ie/aschlote/twinLIN.mov`.

13.2.2 Pollution Modeling and Simulation

In principle a practical application of a cooperative pollution control program would rely upon emissions measurements in a road network. For the sake of experimental validation of the idea, here we adopt an average speed emission model to roughly estimate emissions from each vehicle (see [20]). The average-speed approach is described in detail in the UK Design Manual for Roads and Bridges (DMRB) [20]. According to these models, the emission factor $f(t,p)$ is computed as

$$f(t,p) = \frac{k}{v}\left(a + bv + cv^3 + dv^3 + ev^4 + fv^5 + gv^6\right), \qquad (13.1)$$

where t denotes the type of vehicle (and depends on fuel, emission standard, category of vehicle, engine power), p denotes the pollutant of interest (e.g. CO, $CO2$, NOx, Benzene), v denotes the average speed of the vehicle, and the parameters a, b, c, d, e, f, g and k depend on both the type of vehicle and the pollutant p under consideration. For the purpose of this work, the values of the parameters are taken from Appendix D, in [20]. For convenience, we report in Table 13.1 the emission factors for CO which have been used later in Section 13.3 (parameters d, e, f and g are zero for the chosen pollutant and classes of vehicles, and have not been reported in the table).

---

[3]PLX Devices Inc., 440 Oakmead Parkway, Sunnyvale, CA 94085, USA. Phone: +1 (408) 7457591. Website: `http://www.plxdevices.com`. Last Accessed July 2017.

TABLE 13.1
CO Emission factors for petrol cars (from [20], page 97)

Fuel	Engine capacity (cc)	Emission standard	Coefficients			Adjustment factor	Valid speed range	
			a	b	c	k	Minimum (km/h)	Maximum (km/h)
Petrol	1400-2000	Euro 1	28.261	1.541	9.7686×10^{-4}	1.000	5	120
Petrol	1400-2000	Euro 2	56.750	-2.2582	2.6667×10^{-2}	1.000	5	140
Petrol	1400-2000	Euro 3	35.040	-1.8003	2.4874×10^{-2}	1.000	5	140
Petrol	1400-2000	Euro 4	22.627	-0.68548	1.4443×10^{-2}	1.000	5	140
Petrol	1400-2000	Euro 5	22.627	-0.68548	1.4443×10^{-2}	0.822	5	140

In (13.1) it is assumed that speeds are measured in km/h and emission factors in g/km. At this point it is probably worth noting that the average-speed model suffers from the drawback that very different vehicle operational behaviors are characterized by the same average speed. More realistic and accurate models can easily be embedded into the simulation environment without changing the qualitative features of the simulation (or the analysis). Also, as we shall see, the key point in our work is the assumption that the pollution levels in the group are an increasing function of the number of vehicles in fully ICE mode.

13.2.3   Mathematical Formulation

Many problems in the intelligent transport context can be formulated as utility optimization problems. These have the form

$$
\begin{cases}
\displaystyle \max_{D_1,\dots,D_n} \quad \sum_{i=1}^{n} f_i(D_i) \\[2em]
\text{subject to} \quad \displaystyle \sum_{i=1}^{n} D_i = C,
\end{cases}
\tag{13.2}
$$

where in our situation n is the number of cars in a geographic area, C is a pollution budget for a given pollutant and D_i is the amount of budget allocated to car i. The individual objective functions f_i map D_i to some measure of utility for the individual. We are going to outline a number of potential utility functions later in Section 13.3.3.

A large number of methods is available to solve utility maximization problems. Some could, for example, be solved by a traffic management center and communicated to the vehicles. However, this approach requires a significant amount of communication, accurate measurements, knowledge of all utility functions and perfect compliance of all vehicles. Furthermore, traffic conditions such as the number of vehicles and the driving speed change over time requiring a re-computation of the optimal allocation in short time intervals. In our work, we thus focus on decentralized and iterative approaches to solve this optimization problem with minimal communication overhead. In what follows, we shall reformulate the above optimization as a control theoretic problem and use a stochastic implementation to implement the control strategy. This shall allow us to implement aggregate emissions control based on measurements available to the infrastructure. We repeat that this formulation of the problem completely removes the need for any dedicated V2I or V2V communication, and only requires that cars are able to listen to broadcast information. This gives rise to a feedback loop of the type depicted in Figure 13.4.

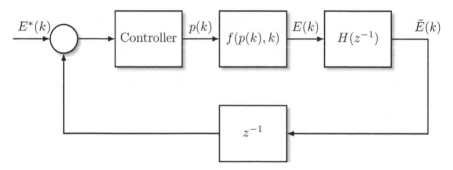

FIGURE 13.4

A basic control loop, with a time-discrete implementation of a classical PI control

13.2.4 Integral Control

In this section we focus on the regulation problem, and wish to maintain pollution in a geographic area below a given threshold by orchestrating the switching of vehicles into fully electric mode. In the next chapter we shall discuss related optimization problems that solve problem (13.2) using the results in [193]. Also, Section 13.3.3 and paper [164] discuss examples of situations where both regulation and optimization are simultaneously achieved.

For the regulation objective here, we consider the following scenario.

(i) Let us denote by $E(k)$ the amount of pollution produced by all vehicles in a geographic area at time instant k. Let $p(k)$ be the probability that an individual vehicle is using its combustion engine at time k. We assume that there is an increasing, possibly non-linear relationship between the average levels of pollution $\bar{E}(k)$ generated by the vehicles (in a geographic area), and the value of the signal $p(k)$.

(ii) Pollution levels in a geographic area can be measured or estimated without explicit communication from the vehicles.

(iii) There is no explicit V2V communication.

(iv) There is no explicit V2I communication, but participating vehicles can listen and respond to broadcast measurements.

With these assumptions we can set up the regulation problem classically as depicted in Figure 13.4 where $E^*(k)$ denotes the desired level of pollution at time instant k. In principle, the feedback control problem can be solved in many different ways, see for instance [164] for a comparison among different control strategies. Here, we simply adopt a PI-like control, where the control action is proportional to the error, and to the integral of the error, between the desired pollution threshold and the actual measured pollution. Roughly speaking, the central entity in Figure 13.3 allows vehicles to travel in ICE mode when pollution is safely below the threshold, and starts broadcasting a probability $p(k)$ smaller than 1 when the safe threshold is exceeded. The more the threshold is exceeded, the smaller is the probability.

13.3 Simulations

13.3.1 Simulation Set-up

This section describes the simulation results obtained by implementing the integral control algorithm described in Section 13.2.4 and explores other more sophisticated scenarios of interest. As a simple example, we consider a grid-like road network simulated in SUMO [113] (i.e., where streets intersect orthogonally), and we measure the aggregate level of pollution using the average model described in Section 13.2.2. Although the investigated grid does not correspond to a particular city, and may be seen as a simplification of a true transportation network, it still reflects the topology of many big cities, that can be found in North America. In order to implement the cooperative pollution control strategies previously described, SUMO data are sampled every 10 seconds, and SUMO is interfaced with MATLAB where the control algorithm was coded.

In the simulation, cars enter from 5 different points of the network and drive along random routes. Over a time interval of 10000 seconds we increase the number of cars in the network until 2000 cars have entered. We then continue the simulation for 6 hours. For the purpose of emissions modeling we assume that all cars are petrol electric hybrid cars with weight below 2.5 Tonnes and with combustion engine capacity between 1400 and 2000 cc, whose emission factors were reported in Table 13.1. We further consider a realistic vehicle mix in terms of emission standards. The evolution of the number of cars over time in the network is depicted in Figure 13.5. To better understand the performance of our algorithms, we simulated what happens if we assume, that all cars decide not to use their electric drive at all, or where all cars are just conventional combustion driven vehicles. Figure 13.5 shows the evolution of emissions over time in this uncontrolled scenario. As one might easily expect, the two quantities are strongly correlated.

13.3.2 Disturbance Rejection

Figure 13.5 illustrated the obvious fact that harmful emissions increase as the number of vehicles in the network increases. Our objective now is to show that using a cooperative pollution control, we can decrease the coupling between the number of vehicles and the aggregate emissions. For this purpose, we deploy the integral control strategy described in Section 13.2.4. We concentrate

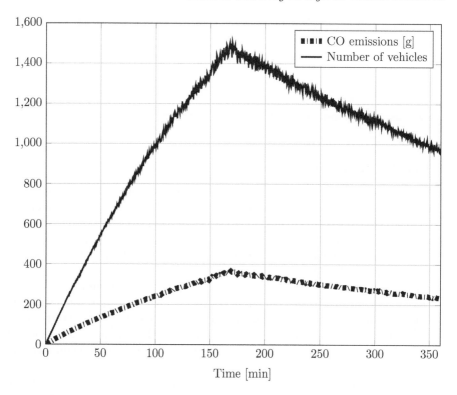

FIGURE 13.5
Evolution of the number of vehicles and corresponding CO emissions in the
network if all cars are combustion driven

here on the control of the carbon monoxide emissions (CO), but the same
approach can be used for any other pollutant or combinations of pollutants.
In particular, we assume that we are interested in regulating the CO emissions to the value 150 grams every minute. The evolution of the emissions and
the corresponding evolution of the broadcast probability in all three scenarios
are depicted in Figures 13.6 and 13.7. Note that the target values we picked
for the pollutants were selected arbitrarily but can easily be adapted to safe
levels.

Next we repeat the above simulations where we also add an external source
of CO that is active from minute 180 to 270 and contributes 40 grams of CO
per minute during this time. In reality this could for example happen if the
wind turns and carries pollution from an industrial area into the city for some
time. In this situation the cars treat this source of pollution as a disturbance
and adjust their behavior to compensate for it. Paradoxically, the aggregate
emissions from the hybrid vehicles can be much cleaner than the background
air quality. The results are depicted in Figure 13.8 where the light curve

shows the pollution caused by PEHVs only, while the darker curve shows that the overall pollution, including the external source as well, is still maintained around the desired value (i.e., the integral control algorithm manages to reject the external disturbance).

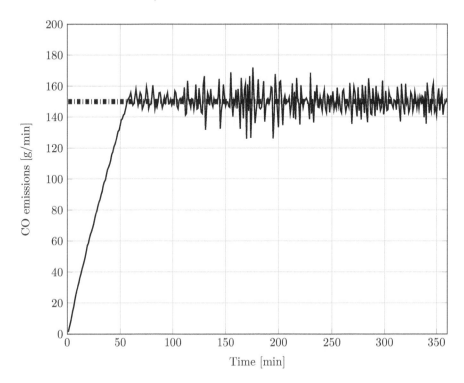

FIGURE 13.6
Evolution of emissions in the controlled case

13.3.3   Extensions

Figures 13.6 and 13.8 clearly emphasize the effectiveness of pollution control strategies in achieving a desired level of pollution. This result was obtained by allowing only some hybrid vehicles to travel in ICE mode, according to a probability distribution that takes into account the distance between the actual pollution level and the desired one (i.e., less hybrid vehicles will travel in ICE mode when the pollution is high). In this sense the broadcast probability can be seen as the desired fraction of cars that on average will travel in ICE mode. In particular, note that the same probability is broadcast to all vehicles.

The previous approach can be greatly enhanced if we allow the central entity to broadcast customized probabilities to each vehicle, rather than the same one to all of them, or if we use the AIMD algorithm, as in Chapter 18. In

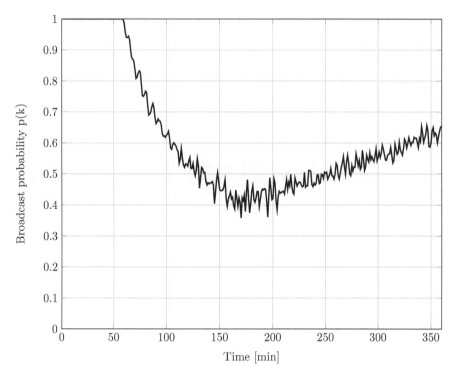

FIGURE 13.7
Evolution of the broadcast probability $p(k)$

this way, other objectives can be achieved as well, in addition to maintaining pollution close to a desired level. In particular, some specific concepts of fairness can be implemented as well, as discussed in Section 13.1. Some of these are discussed in [164] and in later chapters.

13.4 Concluding Remarks

In this chapter we have described how hybrid vehicles can be used to control aggregate traffic related emissions in urban scenarios. Roughly speaking, the idea is to individually control the switching from one driving mode to another of PHEVs traveling in a given area, so that the aggregate emissions do not exceed a desired threshold. For this purpose, as specially modified vehicle was described and demonstrated.

The ideas presented here can be generalized to handle a number of derivative scenarios. For example, it is possible to relax the assumption that all

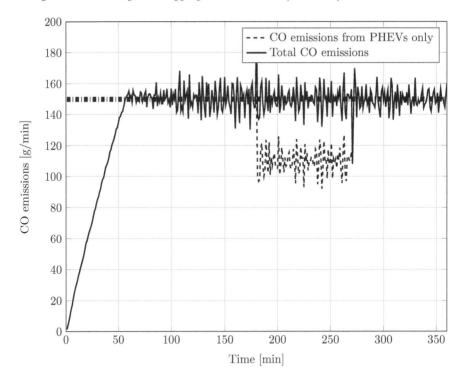

FIGURE 13.8
Pollution is maintained around the desired value, even in the presence of an external pollution source. Pollution caused by single PHEVs is shown with a lighter color

cars are within a small geographic area and, instead, assume that we regard a company that has a fleet of vehicles, such as a taxi company, a delivery company, or a public transport organization. Then, we can give the company an emissions budget that then must be allocated to its vehicles in the same fashion as earlier. Some ideas in this direction are further explored in Chapter 14.

14

Smart Procurement of Naturally Generated Energy (SPONGE)

In most developed economies, road vehicles account for a huge amount of the energy that is consumed by society. The inefficiency of such vehicles has long been the subject of research in the automotive community. At this present point in time, it can be said that three forms of response towards reducing road traffic's thirst for energy have been provided by car manufacturers. The first response was to make vehicles more efficient; both in terms of having better engines ([105], [159], [202]) and improved aerodynamics[1]. This approach has resulted in reductions in energy consumption and research in this direction continues in the area of improving engine combustion efficiency, power-train friction reduction, waste-heat recovery, vehicle rolling resistance, air drag reduction, and improved controls[2]. As a second level of response, automotive manufacturers have also tried to teach people how to drive their vehicles more efficiently. Examples of work in this direction include the ECOWILL project that ran from March 2010 to April 2013, funded by the Intelligent Energy Europe program of the European Union, which aimed to boost and train people across thirteen countries in Europe in eco-driving[3]. Indeed, many vehicles now come with eco-driving options as standard[4], and new vehicle technologies have been widely adopted to alert drivers to more energy efficient driving practices. For instance, gear shift indicators for certain classes of vehicles were recently made mandatory within the European Union[5]. The purpose of the indicators is to inform drivers of when to change gear to minimize fuel consumption. In addition to technological improvements, other instruments have been used to teach environmentally friendly driving behaviors. For example, the project TEAM[6], a European research project co-funded by the European

---

[1]Sustainable Energy Ireland - A Guide to Vehicle Aerodynamics. Available online at http://www.seai.ie/Your_Business/Technologies/Transport/Aerodynamics_Transport_Guide.pdf. Last Accessed July 2017.

[2]http://www.volvobuses.com/SiteCollectionDocuments/VBC/Downloads/Volvo-B8R-Euro6-Brochure-EN.pdf. Last Accessed July 2017.

[3]Project Website: https://ec.europa.eu/energy/intelligent/projects/en/projects/ecowill. Last Accessed July 2017.

[4]https://www.renault.co.uk/discover-renault/innovation-and-technology/eco-driving.html. Last Accessed July 2017.

[5]EUROPA EUR-Lex: Access to European Union law. http://eur-lex.europa.eu/legal-content/EN/TXT/?uri=URISERV:mi0053. Last Accessed July 2017.

[6]Project Website: http://www.collaborative-team.eu. Last Accessed July 2017.

Union, has as an objective, among other applications, to develop a Green, Safe and Collaborative Driving Serious Game and Community Building (SG-CB) application. The aim of this application is to create a gamified social network environment for participants (drivers and travelers) to be able to exchange simple feedback about their current level of performance and thus build a community and together reach higher levels of green driving and lower traffic. Efforts have also been made to encourage car sharing[7],[8]. According to [56], car sharing can reduce car ownership at a rate of one rental car replacing fifteen owned vehicles. The third level of response offered towards reducing vehicular energy consumption concerns the emergence of collaborative and connected vehicle technology. For example, see [122], where a distributed and privacy-aware speed advisory system was proposed in which the objective was to recommend optimal speeds for groups of vehicles traveling along highways to minimize emissions over the entire fleet. In [120], an approach was proposed that simultaneously optimized the numbers and locations of, and speed limits posted on, variable message signs as a means to improve the smoothness and reduce the environmental impact of freeway traffic. Future intended improvements include the ability to incorporate real-time traffic data. In [90], a type of observed collective behavior regarding diverse sets of vehicles traveling along a two-lane highway with different velocities was described.

In this current chapter, we go one step further and report a more holistic view of the energy consumption process and thereby introduce a fourth level of response. Specifically, we wish to allow drivers to make use of (potentially zero cost - free) renewable energy as it becomes available. We do that by allowing weather forecasts to influence the energy management system of vehicles individually and fleet-wise. This provides us with the advantage of using free renewable energy in an optimal way to maximize the efficiency of the transportation fleet. We call this idea Smart Procurement Of Naturally Generated Energy (SPONGE). As we will see, SPONGE can indeed be viewed as an evolution of eco-driving, where we now prime cars to use renewables as they become available, thus following a "use it or lose it" line of thought. Note that, from the perspective of plug-in EVs, the SPONGE solution has also the potential to simplify the "charging paradigm", and to provide a platform for creating aggregated super batteries. Hitherto, most charging research has focused on how to share the available energy among the connected fleet of vehicles in a manner that is compliant with the desires of the EV owners, the constraints of the grid, and the available power. Note that in this case, there might arise some problems in the power grid to accept the unexpected load, with the ultimate possibility of causing thermal overload of network components, low voltages at sensitive locations of the network, and increased phase unbalance ([48]). Even ignoring this, the required optimizations often place severe constraints on the EV owners in the form of inconvenient charging pro-

---

[7]https://www.carsharing.ie. Last Accessed July 2017.
[8]https://www.gocar.ie. Last Accessed July 2017.

files. On the other hand, in a SPONGE context, one would compute the same quantity in advance, and deplete the batteries of the vehicles while traveling of the same quantity (one can even monetize some of this energy to power secondary services from the vehicle such as WiFi or street sensing). Thus, the charging process becomes fully schedulable and programmable. The charging problem can be thus be reduced to a best-effort problem where the cars share the available energy during the charging period using some simple algorithm such as AIMD algorithms ([173] and [43]). Thus, clearly, the difficulties of matching the demand and the offer are shifted to the driving stage through an optimal orchestration of the ICE and the electric engine.

This chapter is organized as follows. Section 14.1 formulates the basic SPONGE problem, in a more formal fashion. Section 14.2 describes the several ingredients that are required to implement the SPONGE approach in practice. Section 14.3 describes a special use case where SPONGE is applied to a fleet of hybrid buses. provides some simulation results to support the efficacy of the proposed methods. Finally, in Section 14.5, we give some simulation results obtained in the use case of the hybrid buses.

14.1 Mathematical Formulation

For convenience, and ease of exposition, we make the following set of simplifying assumptions.

(i) We assume that during some fixed time of the day (e.g., 8am to 6pm), a group of PHEVs will participate in the SPONGE scheme by proactively adjusting their energy consumption patterns at every available clock period to make space available in the battery of the PHEV. We assume that this fixed time period consists of M clock periods and we index every available clock period as $k \in \{1, 2, ..., M\}$.

(ii) We assume that during some other fixed time period of the day, these vehicles are plugged in for charging, and that for this period, a reliable day-ahead forecast of available renewable energy is available. We denote this expected available energy by E_{av}. For example, a typical assumption might be that vehicles charge from 11pm to 6am (i.e., at night time), though it is not necessary for this time period to be the same for all vehicles. Although, in principle, the future horizon of optimization can be longer than one day, weather forecasts might not be reliable enough to support optimal decisions over longer time periods, see ([92] and [203]).

(iii) Each vehicle is assumed to be capable of operating in full electric mode, in ICE mode, or in a combination of both. We have already described in Section 13.2.1 how this could be done, for instance, for the specific case of the Toyota Prius.

(iv) We assume that at any clock period $n(k) \leq N$ vehicles are in transit, and that these vehicles can report their energy consumption over some period to a central agent. In practice, this allows us to regulate energy consumption in a feedback loop, in the same way in which pollution was regulated in Chapter 13, as shown in Figure 13.3.

The SPONGE idea can be mathematically formulated in a manner that is again very similar to the pollution regulation control problem (13.2):

$$\begin{cases} \max_{D_1,\dots,D_n} & \sum_{i=1}^{n(k)} f_i(D_i(k)) \\[2em] \text{subject to} & \sum_{i=1}^{n(k)} D_i(k) = \frac{E_{\text{av}}}{M} \end{cases} \tag{14.1}$$

The optimization problem (14.1) is similar to (13.2), but now the quantity $D_i(k)$ represents a budget of energy expended (i.e., consumed) by the ith vehicle in the interval of time index k, and not a quantity of emissions. In addition, the equality constraint says that the overall amount of energy consumed by all the vehicles $n(k)$ driving on the road in the kth interval should match a pre-fixed quantity E_{av}/M. In particular, the matching between the expected available energy E_{av} and the consumed energy is performed at the kth window of time (i.e., to take into account that different vehicles are on the road during a different time of the day). In particular, note that the main difference between the mathematical formulation of SPONGE (i.e., (14.1)) and the pollution regulation problem described in Chapter 13 is that in the latter case we are interested in regulating the instantaneous value of the emissions. On the other hand, here we are interested in regulating the cumulative value of the energy depleted during a driving period (e.g., during the day).

The optimization problem may be solved in many ways under suitable assumptions on the utility functions f_i. The problem is most interesting when the f_i represent a generalized notion of benefit and this is considered to be a private information, not to be revealed to other vehicles. The problem is then to solve the utility optimization problem in a privacy preserving manner. Note that the f_i may be incorporated to represent various use cases. For example, OEMs may partner with electrical utility companies to provide a service where the price of energy is part of the purchase plans for PHEVs. The owners paying more upfront, may have prioritized access to "free energy" as it becomes available. Alternatively, the f_i could represent the price paid by an individual vehicle owner for energy access. Some hybrid modes blend the EV motor with the ICE to optimize fuel economy and emissions, so another interesting embodiment of the optimization scenario is to take the required energy in a manner that minimizes the impact on fuel economy of the fleet. In this situation, the utility functions could represent the driving efficiency of the driver.

14.2 Practical Implementation

We now briefly comment on the initial implementation and validation of the SPONGE ideas.

A. Large-Scale Traffic Simulator : The fleet of vehicles implementing SPONGE ideas has been simulated in the SUMO environment, using the "remote control" interface Traffic Control Interface (TraCI) [189], that allows one to adapt the simulation and to control singular vehicles on the fly. A full description of an embedding platform, where real vehicle behaviour, and simulations, can be merged, is given in Part IV.

B. Test Vehicle : Validation of the SPONGE concept was carried out using a test vehicle, a 2015 Toyota Prius VVTi 1.8 5DR CVT Plug-in Hybrid vehicle, already illustrated in Section 13.2.1 for the distributed emission control task.

C. Weather forecasting : An important component in any real practical implementation of the SPONGE program is the ability to have a reasonably accurate, and cheap, prediction of the expected energy E_{av} that will be later available for charging. To obtain a feeling for fidelity of such tools, we evaluated the accuracy of a *free* online forecasting tool over a 3 month period. The tool that we evaluated is provided by the Technical University of Crete[9], where the energy generated by a solar plant can be predicted (anywhere in the world) by simply providing the technical parameters of the plant. We collected real data on-site from Photovoltaics (PV) panels mounted on the flat roof of the building in University College Dublin, Ireland. We recorded a total of 100 days and the predicted and the actual recorded energies are shown in Figure 14.1. As also shown in Figure 14.2 the predictions are relatively accurate with 80% of the predictions within 3% of Normalized Mean Absolute Error (NMAE) and the maximum NMAE is 7%. Thus, our data suggests that accurate predictions can be performed even for small powers, and even when a free online tool is employed. As for wind power

---

[9]http://www.intelligence.tuc.gr/renes/. Last Accessed July 2017.

forecasts, we note that a recent study in Germany reported that "typical wind-forecast errors for representative wind power forecasts for a single wind project are 10% − 15% root mean square error of installed wind capacity but can drop down to 6% − 8% for day-ahead wind forecasts for a single control area and to 5% − 7% for day-ahead wind forecasts for all of Germany"[10]. The accuracy may further be increased if other (commercial) tools are employed. From the previous discussion it appears reasonable to claim that on average the prediction error is below 10%, and this is consistent with other recent studies as well, see for instance [92] and [203].

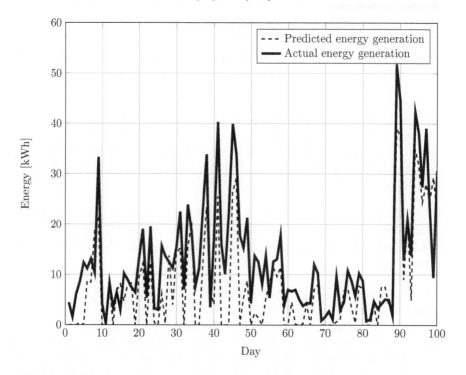

FIGURE 14.1
Comparison between the real and the predicted energy generated from PV panels in UCD

Remark 14.1. While the effect of uncertainty is beyond the scope of the present discussion, we note briefly that it is simple to accommodate for forecasting errors by buying extra energy, if required, from the outer grid, or by appropriately using other storage devices, if available. However, interactions with the grid are not always convenient, either in terms of price, or in terms of environmental friendliness of the average power mix from the grid (see [175]).

---

[10]https://www.nrel.gov/grid/solar-wind-forecasting.html. Last Accessed July 2017.

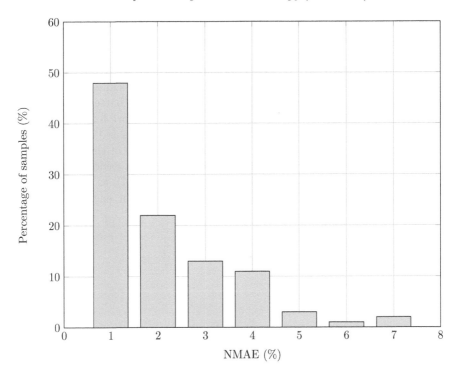

FIGURE 14.2
Histogram of the percentage of NMAE

An alternative to this is to formulate an uncertainty description as part of the optimization.

14.2.1   SPONGE Simulation Results

We now briefly report some simulation results to show a possible solution of the SPONGE optimization problem formulated in Equation (14.1). As previously stated, this problem is regulatory in nature, the idea being to regulate a fleet-wise electric energy consumption. The kinds of signals broadcast by the central authority to orchestrate fleet behavior are probabilistic in nature, where the probability that vehicles are directed to travel in EV mode is a function of the gap between the desired target of electric energy that the fleet should consume, and the energy that each vehicle has already consumed. The objective is to achieve an overall energy consumption equal to the expected energy that will be available from renewable sources in the next charging period. In particular, assume that 40 PHEVs participate to the SPONGE scheme from 9:00 to 18:00. The iteration step-size of the algorithm is assumed to be 1 minute, thus providing a reasonable switching time interval for each PHEV operating in either EV or ICE mode. Accordingly, the algorithm is iterated 540 times overall. In addition, assume that the total available renewable energy for the next charging period is equal to 540 kWh, so that energy should be allocated with a rate of 1 kWh/minute from 9:00 to 18:00, to match the final target. Finally, each PHEV is assumed to have the same battery capacity of 20 kWh with the initial state of charge (SOC) of the battery equal to 80%. SUMO, and its "remote control" interface, TraCI (short for Traffic Control Interface) [6], is used to simulate the motion of vehicles and control the driving mode of single vehicles on the fly. For the purpose of this simulation, as in [80], we shall consider that vehicles drive in the area enclosing the campus of the National University of Ireland Maynooth (NUIM), as shown in Figure 14.3. At every time step, each PHEV sends its current energy state to the

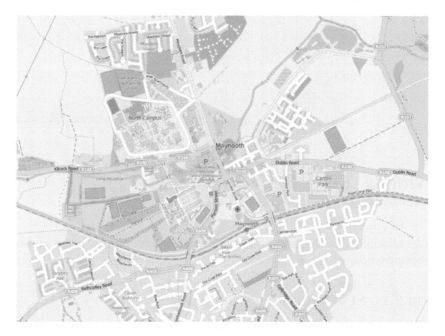

FIGURE 14.3
Area near the NUIM campus, used for our simulation set-up

central infrastructure. Upon receiving these data, the central infrastructure calculates and broadcasts some global signals to all PHEVs. Upon receiving this information, each PHEV updates its probability to travel in EV mode. Then, the PHEV compares such a value with its own "coin-flipped" value, which is a uniformly distributed random number in the range [0, 1], and finally decides which mode (i.e., EV or ICE) the vehicle should travel during the next time interval, using a similar strategy to the one described in Chapter 13. Note that it is assumed here that the electric mode of each PHEV can only have two states, i.e., either completely on (value 1) or off (ICE mode, value

0), during each unit of time (i.e., 1 minute). In the SUMO simulations, the practical energy consumption of each vehicle was calculated according to an approximated linear mapping between the traveled distance and the SOC. The simulation results are illustrated in Figures 14.4 and 14.5. In Figure 14.4, it is shown that the overall energy consumption of the PHEVs is indeed controlled to eventually achieve the expected energy target. As a further result, Figure 14.5 shows that the final objective is achieved by maintaining the energy rate constant throughout the day (i.e., around 1 kWh/minute).

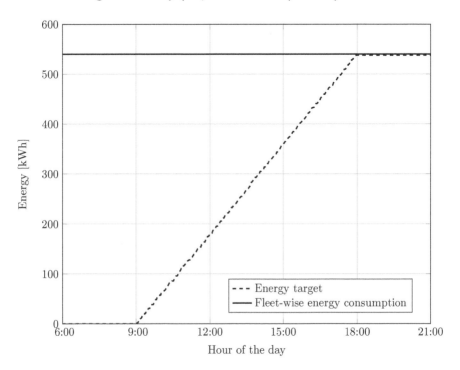

FIGURE 14.4
Evolution of the cumulative target energy allocated to the fleet of PHEVs

Remark 14.2. As a general remark, note that deploying the algorithm of *twin-LIN* in the SPONGE case presents two main drawbacks: (i) there is a continuous communication between the infrastructure and the vehicles (i.e., to continuously communicate the level of energy); and (ii) the gains on the control algorithm depend on the dimension on the network (i.e., the number of vehicles participating to the SPONGE program). Accordingly, in the remainder of this chapter we shall describe a different algorithm to apply SPONGE-like ideas, in a specific case study, to overcome the previous drawbacks.

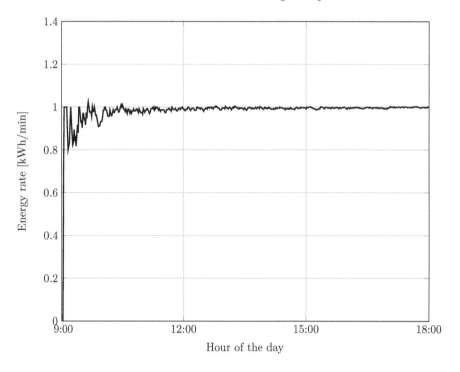

FIGURE 14.5
Evolution of the averaged allocation rate of target energy to the fleet of PHEVs

14.3   Specific Use Case: SPONGE for Plug-in Buses

While in principle the SPONGE program can be applied to regulate energy consumption in any fleet of PHEVs, a special case arises when the fleet consists of Plug-in Hybrid Electric Buses (PHEBs). PHEBs are increasingly seen as an effective tool in combating air pollution in some cities, and as a tool for reducing cities' reliance on fossil fuels (thereby reducing greenhouse gas emissions) [127, 97]. Consequently, the design and operation of such buses has been the subject of much research interest. Hitherto, significant research effort has focused on improving the fuel economy while guaranteeing that both the ICE and the electric machine work in the high-efficiency area; typically, by taking into account knowledge of both bus routes and passenger loadings in a predictive manner. Selected examples of work in this direction can be found in [182, 117, 116]. The objective now is to extend this line of inquiry further, and use the degree of freedom of their electric motors to further accommodate energy from renewable sources, as in the SPONGE spirit.

14.3.1 Sponge Bus Problem Formulation

Let $i = \{1, 2, ..., N\}$ index the set of N PHEBs participating to the SPONGE program. We shall make the following assumptions:

- It is assumed that after a number of trips along their (different) routes, the N PHEBs stop for charging at the bus station. For instance, we can assume that the PHEBs will not drive from 11pm to 6am, and they will be charged in this time frame;

- It is also assumed that a 24-hour ahead forecast of energy from the renewable energy sources will be available (e.g., a forecast of how much energy will be generated by the wind plants connected with the charging station at night time). We denote this amount of energy available by E_{av} as before;

- Early in the morning, before being dispatched along their routes, the buses will compute how the energy E_{av} should be optimally shared among themselves during the day (i.e., in terms of energy consumption of their own batteries);

- In order to compute the optimal allocations of energy, we shall assume that each PHEB is equipped with a device to transmit messages to the central infrastructure via Vehicle to Infrastructure (V2I) technology;

- The central infrastructure has the ability to broadcast messages to the whole network of PHEBs using some Infrastructure to Vehicle (I2V) technology.

Note that it is not required for vehicles to exchange information among themselves, and thus, it is not required PHEBs for the vehicles to be equipped with V2V communication devices. When traveling along their routes, the buses will be able to choose when it is more convenient to switch from electric mode to ICE mode (i.e., using the ICE) and back. In this context, D_i denotes the energy consumption by the ith bus along its trip. Then, we are interested in computing the solution of the following optimization:

$$
\begin{cases}
\displaystyle \max_{D_1, D_2, ..., D_N} \ \sum_{i=1}^{N} f_i(D_i) \\[2mm]
\text{s.t.} \ \displaystyle \sum_{i=1}^{N} D_i = E_{av}
\end{cases}
. \tag{14.2}
$$

In the optimization problem (14.2), the terms D_i can be interpreted as a "budget" of energy that is allocated to the ith bus in order to maximize a utility function of interest, such that the sum of the energy budgets allocated to all the buses matches E_{av} as in the SPONGE spirit. Although the mathematical formulation (14.2) closely resembles that of (14.1) in the case of cars, still there are two main differences that should be kept in mind when solving the problem:

1. The route of buses is known in advance. Accordingly, it is possible to take it directly into account in the objective function.

2. Buses of the same company may be expected to cooperate to achieve a common goal. In principle, on the other hand, cars' owners are expected to behave in a selfish way to achieve a personal goal.

In this work we shall explore the particular case where one is interested in maximizing the saving of CO_2 emissions. Clearly, each f_i is an increasing function of D_i as no CO_2 emissions are saved when the bus travels all the time in ICE mode, while no pollution occurs when all the vehicles travel in electric mode all the time. The utility functions of 15 PHEBs that we shall study are shown in Figure 14.6 as a function of the percentage of the use of the electrical engine for each bus. These functions are constructed from real data and the next section will explain how the utility functions are designed in detail.

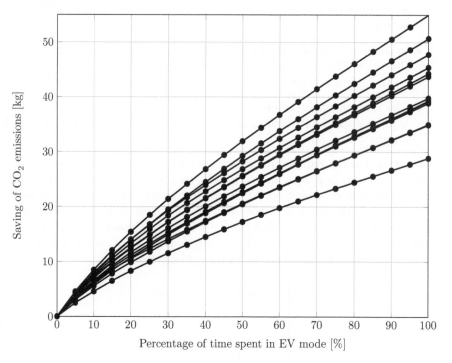

FIGURE 14.6
Utility functions of 16 PHEBs in Dublin city. Note that some buses pollute more than others (and thus, have a greater potential in terms of CO_2 savings) depending on the characteristics of their routes (e.g., speed limits). The dot' points mark the fitted curves of utility functions using cubic splines.

14.3.2   Construction of the Utility Functions

14.3.2.1   Electrical Energy Consumption

Under the assumption that a vehicle is traveling at a constant velocity s, the amount of electrical energy consumption of vehicles can be modeled as a convex function of s, see [185] for instance. The convex function depends in turn on, among other things, the physical characteristics of the bus. In our work, we used the real energy consumption data of a BYD electric bus[11] and noticed that it can be accurately approximated with a polynomial function of degree 4 of the vehicle speed s as

$$e(s) = \alpha_0 s^4 + \alpha_1 s^3 + \alpha_2 s^2 + \alpha_3 s + \alpha_4. \tag{14.3}$$

where $\alpha_0, \alpha_1, \alpha_2, \alpha_3, \alpha_4$ are all constant parameters. Using a conventional least square method to fit the real energy consumption, the following values for the parameters were obtained in [137]

$$\alpha_0 = 3.1970 \cdot 10^{-4},$$
$$\alpha_1 = -0.0604,$$
$$\alpha_2 = 4.3123,$$
$$\alpha_3 = -151.2257,$$
$$\alpha_4 = 3.3288 \cdot 10^3,$$

and the corresponding utility function is depicted in Figure 14.7 which shows both the empirically measured data and the fitted polynomial function. Although energy consumption is known to increase with the cube of the speed for aerodynamic reasons, still, it is very large for low speeds as well, because the energy required for ancillary services (e.g., air conditioning) increases when traveling time increases. This aspect has been extensively discussed in [185], among others, and is consistent with collected experimental data[11].

Remark 14.3. While we acknowledge that (14.3) provides only a rough estimate of how much energy may be consumed by a bus along a stretch of its route, still it provides a proxy of the actual quantity which may be enough for our forecasting purposes.

14.3.2.2   Saving of CO_2

In an analogous manner to power consumption, CO_2 emissions may also be computed as a function of the speed of the vehicles, by adopting for instance the average-speed model from [20], already described in Section 13.2.2. In particular, here we use function

$$h(s) = k \left(\frac{a + bs_i + cs^2 + ds^3 + es^4 + fs^5 + gs^6}{s} \right), \tag{14.4}$$

---

[11]http://insideevs.com/byd-electric-bus-test-results-in-canada/. Last Accessed July 2017.

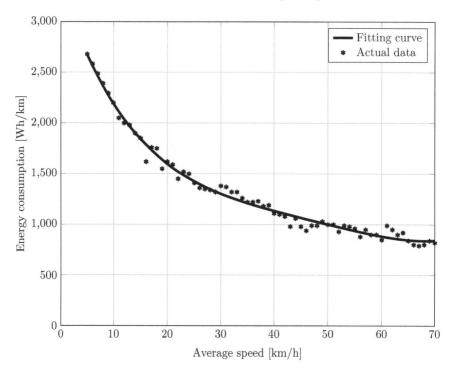

FIGURE 14.7

A typical energy cost function for PHEBs traveling at a constant speed

where $a, b, c, d, e, f, g, k \in \mathbb{R}$ are chosen corresponding to the vehicle code R203 in [20] (i.e., diesel buses with up to 15 t of gross vehicle mass).

14.3.2.3   Utility Functions f_i

The overall utility function f_i quantifies how much CO_2 has been saved by the ith bus, provided that the bus is allowed to spend a budget of D_i units of energy when traveling along its route. In the following, we shall assume that the whole path traveled by a bus during the day can be split into a number of very small sections, corresponding to the distance traveled by a bus in one second. For simplicity, we shall assume that the bus speed in each of the sections is constant and is a proper fraction of the posted speed limit to take into account the possible effects of congestion events. The speeds will then be used in the computation of the CO_2 emissions (14.4) and the energy usage (14.3). Also, we shall denote by $\mathcal{R}_i$ the set of all the sections traveled by the ith PHEB, and by γ_l the fraction of the time that a PHEB will travels in EV mode along the lth section of the route. Note that $\mathcal{R}_i$ is the set of all sections traveled by a bus during a day (or in general, between two different runs of the optimization algorithm). Given that bus routes are typically cyclic, this implies that the same stretch of a road might appear more times in $\mathcal{R}_i$, and possibly with different values of optimal γ_l. To construct the utility functions, we make the assumption that every PHEB will optimally use the allocated energy D_i and thus the corresponding utility function $f_i(D_i)$ for PHEB i is given by

$$
\begin{cases}
f_i(D_i) = \max_{\gamma_l} \quad \sum_{l \in \mathcal{R}_i} h\left(s(l)\right) \cdot \gamma_l \\[2ex]
\text{s.t.} \quad \sum_{l \in \mathcal{R}_i} e(s(l)) \cdot L(l) \cdot \gamma_l = D_i \\[2ex]
0 \le \gamma_l \le 1, l \in \mathcal{R}_i
\end{cases}
\tag{14.5}
$$

where $L(l)$ denotes the length of the lth section of a trip. Due to the fact that all bus routes are fixed and known a priori, given a fixed D_i, (14.5) is a linear program with a single budget constraint (i.e., a continuous linear knapsack problem [45]) and thus the optimal electric energy allocation can be easily computed by sorting the trip sections by decreasing order of the ratio $\frac{h(s(l))}{e(s(l))}$ and then activating the electric engine according to the sorted order. The utility function of each PHEB can thus be computed off-line. Particularly, for each bus i, we vary D_i between 1 and 100 in steps of 1 and compute the optimal $f_i(D_i)$. We note that (14.5) is a parametric linear program with parameter D_i, and thus $f_i(D_i)$ for all $i \in \mathcal{N}$ is a piece-wise concave function [4]. However, since the derivative $f_i'(D_i)$ may be required by the optimization algorithm, as described in more detail in the next section, we have approximated $f_i(D_i)$ for each bus by using cubic spline functions. The resulting utility functions for the 16 PHEBs that are used in the illustrative example of this paper are shown in Figure 14.6.

14.4   Optimization Problem

In principle, many different methods may be used to solve the optimization problem (14.5) to compute the optimal energy budgets D_i for each bus. In particular, in Section 14.2.1 we had already explained a possible solution via a PI-like control, in the context of the distributed regulation of pollution. Now we shall see an alternative solution based on the AIMD algorithm used in earlier chapters. Note that we now use the AIMD algorithm to solve optimization problems (see Part IV). This is particularly convenient for this specific case, as it allows us to simply embed the utility functions of single buses. In particular, this choice is motivated by the following main reasons:

- **Low-communication requirements:** Although we have presented here a simple case study with a small number of buses, the same program can be easily generalized to include hundreds of buses. Also, the batch optimization formulation (i.e., before the buses are dispatched during the day) might be actually solved in real-time to account for non fully-predictable aspects (for example to respond to traffic peaks or weather forecast updates). In this context, it is convenient to consider the communication cost of solving the optimization algorithm. *AIMD based optimization can be solved using only intermittent binary feedback and can thus, unlike many other distributed optimization techniques, be solved without the need to broadcast the Lagrange multipliers in a pseudo-continuous manner.*

- **Privacy-preservation requirements:** In our application, the utility functions f_i potentially reveal sensitive private information. For example, when formulated in a slightly different manner, these functions may reveal how good a particular driver is on a given route. This information is potentially very useful for an employer and could potentially be used in a nefarious manner. In addition, in unionized environments, revealing these functions to an employer could also be of concern and consequently impede the adaptation of ideas like SPONGE. Given this context, a natural question is whether the distributed optimization can be solved without revealing private information, and in this regard *AIMD has some very nice privacy properties.*

- **Agent actuation:** *AIMD requires very little actuation ability on the agent-side.* This is in contrast to other distributed algorithms (like the PI-like control or ADMM (see [21]) where at each time step, agents must solve a local optimization problem.

- **Algorithm parameterization:** *In AIMD the gain parameters of the network are independent of network dimension; rather, they only depend on the largest derivative over all utility functions.* Thus, selecting a gain for the algorithm is extremely simple in the case of AIMD.

A pseudo-code to implement the AIMD algorithm for this specific application is given in Algorithm 14.1, while a more detailed discussion of the AIMD algorithm and its properties is provided in Chapter 18.

Note that the algorithm does not compute the optimal budgets D_i in a single step, but in an iterative fashion, as $D_i(k)$ represents the value of the unknown energy to be allocated to the ith PHEB, computed at time step k. In Algorithm 14.1, k_{max} represents the maximum number of iterations before the algorithm stops (e.g., after five minutes of iterations). The basic idea of Algorithm 14.1 is that if the sum of the $D_i(k)$ of all PHEBs is smaller than E_{av}, then each PHEV increases its target energy consumption $D_i(k)$ at the next iteration $k+1$ by a quantity α. However, if the sum of the energy budgets of all PHEBs exceeds E_{av} (this situation is usually called a congestion event), then

---

Algorithm 14.1 Unsynchronized AIMD Algorithm

1: **Initialization:** $k = 1$, $D_i(k) = 0$;
2: Broadcast the parameter Γ to the entire networks;
3: **while** $k < k_{\max}$ **do**
4:      **if** $\sum_{i=1}^{N} D_i(k) < E_{\mathrm{av}}$ **then**
5:          $D_i(k+1) = D_i(k) + \alpha$
6:          **else** with probability $p_i(k) = \Gamma \frac{1}{\overline{D_i}(k) f_i'(\overline{D_i}(k))}$, $D_i(k+1) = \beta D_i(k)$
7:          **else** $D_i(k+1) = D_i(k) + \alpha$
8:      **end if**
9:      $k = k + 1$
10: **end while**

---

each PHEB decreases its energy consumption by a multiplicative factor $0 < \beta < 1$ with probability $p_i(k) = \Gamma \frac{1}{\overline{D_i}(k) f_i'(\overline{D_i}(k))}$, where Γ is a constant common broadcast parameter, and $\overline{D_i}(k)$ is the time average of the sequence of $D_i(k)$ at congestion events, up to the last iteration. The motivation for this algorithm is as in [193]. Here the p_i's are chosen to be inversely proportional to the derivatives of the utility functions to ensure that they are strictly increasing (a key ingredient of the proof in [193]) for the concave utility functions used here; with the optimal point being a maximum (the f_i's are concave here) as opposed to a minimum in [193].

14.5 Simulation Results

We assume that 16 PHEBs participate to a SPONGE program in Dublin city, Ireland. The area of interest is depicted in Figure 14.8, and has been imported from OpenStreetMap [84] into SUMO. We further assume that weather forecasting tools predict an availability of 500 kWh in the next charging period. Ten minutes before starting their routes, the buses and the CA solve the optimization problem using the described AIMD algorithm, and optimally allocate the 500 kWh of available energy to the 16 different buses. In particular, Figure 14.9 compares the energy that would be required by each bus to travel all the time in EV mode (white bar) with the actually available one, as optimally allocated by the AIMD algorithm (black bar). Finally, Figure 14.10 shows that the AIMD algorithm converges to the optimal solution that can be easily computed by solving (14.2) using a full-information centralized optimization solver.

FIGURE 14.8
Road network of Dublin City, Ireland, imported from OpenStreetMap, used in our simulations

14.6   Concluding Remarks

In this chapter, we have described a strategy that takes advantage of the ability of PHEVs to travel in both the electric and fuel modes to absorb naturally generated electrical energy in a smart manner from the grid. Then we have tailored this idea to the special case of a fleet of PHEBs that follow different routes with different energy requirements.

    The work can be extended in a number of directions. In particular, it might be useful to take into account the driver behavior as a further input into the

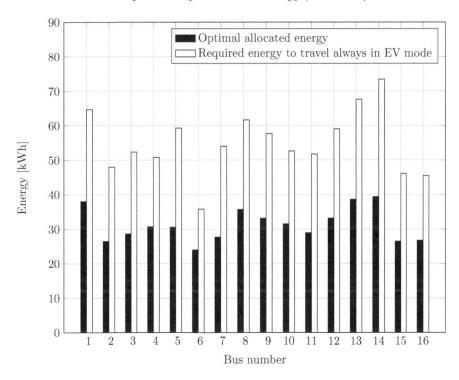

FIGURE 14.9
Comparison of the energy required to travel all the time in EV mode, and the optimal energy for the 16 buses

design of the utility functions. Also, a seamless integration of the proposed idea into the hybrid drive cycle of the vehicles is a further interesting aspect.

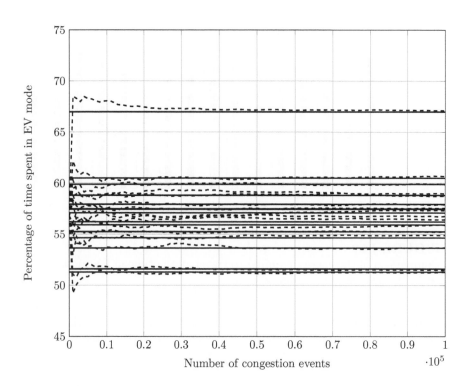

FIGURE 14.10
The AIMD solution converges to the optimal solution obtained by solving the
optimization problem in a centralized, full-information, way

15

An Energy-Efficient Speed Advisory System for EVs

15.1 Introduction

Intelligent Speed Adaptation (ISA) systems, often embedded as a component of Advanced Driver Assistance Systems (ADASs), have become a fundamental part of Intelligent Transportation Systems (ITS). They can often result in improved vehicle and pedestrian safety, a better utilization of the road network, and reduced emissions. In the literature, many papers have addressed their design from different viewpoints, including the perspective of road operators, infrastructure providers and transportation engineers, and some interesting results have been illustrated among others in ([72, 198, 26, 32, 3, 95, 184]).

In this chapter we are interested in designing a speed advisory problem for a fleet of electric vehicles. Such a situation might occur in some sensitive areas in city centers that are closed to conventional traffic, and only allow transit to specific categories of low (or zero) polluting vehicles, see for instance the case of *Umweltzonen*[1] in Germany; or a similar problem might occur for a fleet of urban electric buses, as investigated in Chapter 14. In particular, the use of electric buses to decrease harmful emissions and noise pollution is becoming widespread in different cities of the world, see for instance the recent cases of São Paulo[2], Louisville in the US[3] or Wien in Europe[4].

The starting point of the work described in this chapter is the observation that, roughly speaking, different EVs are designed to operate optimally (e.g., in terms of energy efficiency) at different vehicle speeds and at different loading conditions. In this chapter, the problem is addressed, of computing the optimal speed that should be recommended to all vehicles belonging to the fleet of EVs, in order to minimize the overall energy consumption of the fleet; or similarly, to extend their range. Clearly, this task is performed provided

---

[1]http://gis.uba.de/website/umweltzonen/umweltzonen.php. Last Accessed July 2017.

[2]https://cities-today.com/sao-paulo-to-introduce-its-first-fleet-of-fully-electric-buses/. Last Accessed July 2017.

[3]http://cleantechnica.com/2015/01/16/louisville-gets-first-fully-electric-zero-emissions-city-buses/. Last Accessed July 2017.

[4]http://www.siemens.com/innovation/en/home/pictures-of-the-future/mobility-and-motors/electric-mobility-electric-buses.html. Last Accessed July 2017.

that some basic safety and QoS requirements are guaranteed (i.e., the optimal recommended speed should be within a reasonable realistic range). As we shall see in the remainder of the chapter, the optimal speed heavily depends on how a single EV travels in traffic (e.g., whether air conditioning is on or off and how many people are on board). Since people might not be interested in sharing such a private piece of information, we are interested in obtaining the optimal solution without requiring single vehicles to communicate personal information to other vehicles or even to a central infrastructure. Accordingly, in our work we adopt a recently proposed algorithm (see [122]) that is a consensus algorithm that somewhat preserves the privacy of individual vehicles, as will be better illustrated later in the chapter.

This chapter is organized as follows: Section 15.2 illustrates a basic model of power consumption in electric vehicles, that is a simplified version of the one used in Chapter 4. Such a model will be required to formulate single utility functions that relate the traveling speed with the energy efficiency. Section 15.3 describes the algorithm proposed in [122] and adapts it to the specific scenario of interest. Section 15.4 reports simulation results to show the efficacy and the performance of the proposed approach, and finally Section 15.5 concludes the chapter and outlines our current lines of research.

15.2 Power Consumption in EVs

Most of the discussion here follows the reference ([185]), where the ranges of EVs are reported for different brands and under different driving cycles. Power consumption in an EV driving at a steady-state speed (along a flat road) is caused by four main sources:

- *Aerodynamics* : aerodynamic power losses are proportional to the cube of the speed of the EV, and depend on other parameters typical of a single vehicle, such as its frontal area and the drag coefficient (which in turn, depends on the shape of the vehicle);

- *Drivetrain* : drivetrain losses result from the process of converting energy in the battery into torque at the wheels of the car. Their computation is not simple, as losses might occur at different levels (in the inverter, in the induction motor, gears, etc.); in some cases, power losses have been modeled as a third-order polynomial, whose parameters have been obtained by fitting some experimental data (see ([185]);

- *Tires* : the power required to overcome the rolling distance depends on the weight of the vehicle (and thus, on the number of passengers as well), and is proportional to the speed of the vehicle;

- *Ancillary systems* : this last category includes all other electrical loads in the vehicle, such as Heating, Ventilation, Air Conditioning (HVAC) systems, external lights, audio, battery cooling systems, etc. In this case the power consumption does not depend on the speed of the vehicle and can be represented by a constant term that clearly depends on external factors (e.g., weather conditions) and personal choices (desired indoor temperature, volume of the radio, etc.). According to experimental evaluations (see again ([185]), the power losses due to ancillary services usually vary between 0.2 and 2.2 kW.

If we sum up all the previous terms, then the power consumption P_{cons} can be represented as a function of the speed v as:

$$\frac{P_{cons}}{v} = \frac{\alpha_0}{v} + \alpha_1 + \alpha_2 v + \alpha_3 v^2 \qquad (15.1)$$

where on the left hand side we have divided the power by the speed, to obtain an indication of energy consumption per kilometer, expressed in kWh/km. Such a unit of measurement is usually employed in energy-efficiency evaluations. Accordingly, Figure 15.1 shows a possible relationship between speed and power consumption, obtained using data from Tesla Roadster (see [185]) and assuming a low power consumption for ancillary services of 0.56 kW (i.e., assuming air conditioning switched off). As can be noted from the figure, there is a large energy consumption when the speed is large (due to the fact that power increases with the cube of the speed for aerodynamic reasons), but also for low speeds, due to the fact that travel times increase, and accordingly constant power required by ancillary services requires more energy than the same services delivered with high speeds.

In order to implement the proposed system, we shall further assume here that vehicles are equipped with a V2X communication system, and can exchange information with other vehicles at a short distance, and with the infrastructure. For instance, each vehicle could be equipped with a specific communication device (e.g., a mobile phone with access to WiFi/3G networks). Note that such a technological equipment, though usually already available in most EVs, will further have an impact on the energy consumption pattern. Also, note that such an assumption had not been considered in Chapter 13 or Chapter 14.

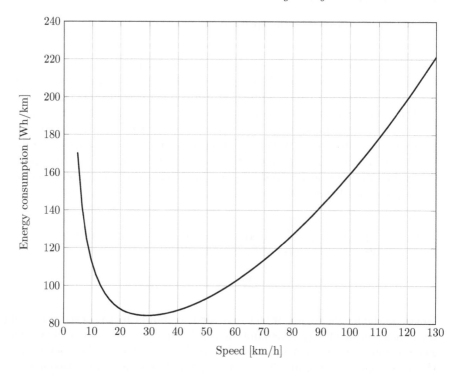

FIGURE 15.1
Energy consumption as a function of the speed of the EV

15.3   Algorithm

Assume that N EVs have access to a common clock (for example, a GPS clock). Let $k \in \{1, 2, 3, ...\}$ be discrete time instants in which new information from vehicles is collected and new speed recommendations are made. Let $s_i(k)$ be the recommended speed of the vehicle $i \in \underline{N} := \{1, 2, ..., N\}$ calculated at time instant k. Thus, the vector of recommended speeds for all vehicles is given by $\mathbf{s}(k)^{\mathrm{T}} := [s_1(k), s_2(k), ..., s_N(k)]$, where the superscript T represents the transposition of the vector. Note that between two consecutive time instants $(k, k + 1)$, the recommended speeds are constant while the driving speeds are time-varying real-valued variables. We denote by N_k^i the set of neighbors of vehicle i at time instant k, i.e., those vehicles which can successfully broadcast their recommended speeds to vehicle i.

In addition, we assume that each vehicle i can evaluate a function f_i that determines its average power consumption at a given steady-state speed, according to (15.1). Note that in order to achieve this, it is necessary that the vehicle both knows its parameters in the function, and also monitors the functioning of some electric appliances on-board (e.g., the intensity of the HVAC system or whether the radio is switched on). We shall assume that functions f_i are convex, continuously differentiable and with a Lipschitz continuous first derivative f_i' which is assumed with positive bounded growth rate in the domain of interest $\mathcal{D}$. We assume that the recommended speed can vary within the domain $\mathcal{D} = [5, 130]$, which is a realistic range of speeds, expressed in km/h. Then, the requirement on the derivative can be expressed as

$$0 < d_{\min}^{(i)} \leq \frac{f_i'(a) - f_i'(b)}{a - b} \leq d_{\max}^{(i)}, \tag{15.2}$$

for all $a, b \in \mathcal{D}$ (i.e., for reasonable steady-state speeds) such that $a \neq b$, and suitable positive constants $d_{\min}^{(i)}$, $d_{\max}^{(i)}$. Notice that (15.1) fulfills all the previous requirements, and thus, the previous assumptions are usually satisfied in the application of interest here. In this context, we consider the following problem:

Problem 1: *Design a speed advisory system system for a fleet of EVs connected via V2X communication systems. The speed recommended by the system is the speed that minimizes the total power consumption of the fleet of vehicles.*

We now formulate the problem as follows:

$$\begin{cases} \min_{\mathbf{s} \in \mathbb{R}^N} & \sum_{j \in \underline{N}} f_j(s_j), \\ \text{s.t. } s_i = s_j, \forall i, j \in \underline{N}. \end{cases} \tag{15.3}$$

This problem is an optimized consensus problem and can be solved in a variety of ways (for example using ADMM [21]). Our focus in this present work is not to construct a fully distributed solution to this problem, but rather to construct a partially distributed solution which allows rapid convergence to the optimum, without requiring the vehicles to exchange information that reveals individual cost functions to other vehicles. This is the privacy preserving component of our problem statement.

Remark 15.1. We shall not address Problem 1 with the objective to calculate the optimal speed to be recommended to all the vehicles in one step. On the other hand, we propose an iterative algorithm that at each step yields individual recommended speeds that will eventually converge to the same value under a consensus constraint. In doing this, we shall assume that the vehicles will be compliant with the recommended speed (this might be more realistic for public transportation rather than for single vehicles, but non-compliance with the recommended speed is not investigated here).

To solve (15.3) we use an iterative feedback scheme of the form

$$s\left(k+1\right) = \mathbf{P}\left(k\right)s\left(k\right) + G\left(s\left(k\right)\right)e, \tag{15.4}$$

where $\{\mathbf{P}\left(k\right)\} \in \mathbb{R}^{N \times N}$ is a sequence of row-stochastic matrices[5], $e \in \mathbb{R}^N$ is the column vector with all entries equal to 1, and $G : \mathbb{R}^N \mapsto \mathbb{R}$ is a continuous function with some assumptions to satisfy as we shall see in (15.9). Similar algorithms were proposed and studied, among others, in [111, 110]; they were further investigated in [122].

Here, we shall assume that (15.3) has a unique solution. Then, according to elementary optimization theory, if all the f_i are strictly convex functions, then the optimization problem (15.3) has a solution if and only if there exists a $y^* \in \mathbb{R}$ satisfying

$$\sum_{j=1}^{N} f_j'(y^*) = 0. \tag{15.5}$$

In this case by strict convexity y^* is unique and the unique optimal point of (15.3) is given by

$$s^* := y^*e \in \mathbb{R}^N. \tag{15.6}$$

In order to obtain convergence of (15.4) we select a feedback signal

$$G\left(s\left(k\right)\right) = -\mu \sum_{j=1}^{N} f_j'\left(s_j\left(k\right)\right). \tag{15.7}$$

and we obtain the dynamical system

$$s(k+1) = \mathbf{P}(k)s(k) - \mu \sum_{j=1}^{N} f_j'(s_j(k))e, \tag{15.8}$$

In [124] it is shown that if $\{\mathbf{P}(k)\}_{k \in \mathbb{N}}$ is a uniformly strongly ergodic sequence[6] and μ is chosen according to

$$0 < \mu < 2 \left(\sum_{j=1}^{N} d_{\max}^{(j)} \right)^{-1}, \tag{15.9}$$

then (15.8) is uniformly globally asymptotically stable at the unique optimal point $\mathbf{s}^* = y^*e$ of (15.3). More details, and the mathematical proofs can be found again in [122].

Thus, we proceed as follows: For each k we define $\mathbf{P}\left(k\right)$ as

$$\mathbf{P}_{i,j}\left(k\right) = \begin{cases} 1 - \sum_{j \in N_k^i} \eta_j, & \text{if } j = i, \\ \eta_j, & \text{if } j \in N_k^i, \\ 0, & \text{otherwise.} \end{cases} \tag{15.10}$$

---

[5]Square matrices with non-negative real entries, and rows summing to 1.
[6]That is, for every $k_0 \in \mathbb{N}$ the sequence $\mathbf{P}(k_0), \mathbf{P}(k_0+1)\mathbf{P}(k_0), \ldots, \mathbf{P}(k_0+\ell)\cdots\mathbf{P}(k_0), \ldots$ converges to a rank one matrix. See [124] for further details.

where i, j are the indices of the entries of the matrix $\mathbf{P}(k)$, and $\eta_j \in \mathbb{R}$ is a weighting factor.

The assumption of uniform strong ergodicity holds if the neighborhood graph associated with the problem has suitable connectivity properties. If sufficiently many cars travel in the city center area, it is reasonable to expect that this graph is strongly connected at most time instances. Weaker assumptions are possible but we do not discuss them here; see [134] for possible assumptions in this context. In any case, note that the time-varying communication graph makes the dynamic system (15.4) become a switching system.

Now, we propose the following algorithm for solving (15.3). The underlying assumption here is that at all time instants all EVs communicate their value $f_j'(s_j(k))$ to the base station, which reports the aggregate sum back to all EVs. This is precisely the privacy preserving aspect of the algorithm, as EVs do not have to reveal their cost functions to the base station, nor to other vehicles. Some implicit information (i.e., the derivative of the cost function at certain speeds) is indeed revealed to the base station but not to the other EVs in the fleet.

Algorithm 15.1 Optimal Decentralized Consensus Algorithm

    for $k = 1, 2, 3, ..$ **do**

        for each $i \in \underline{N}$ **do**

            Get $\tilde{F}(k) = \sum\limits_{j \in \underline{N}} f_j'(s_j(k))$ from the base station.

            Get $s_j(k)$ from all neighbors $j \in N_k^i$.

            Do $q_i(k) = \eta_i \cdot \sum\limits_{j \in N_k^i} (s_j(k) - s_i(k))$.

            Do $s_i(k+1) = s_i(k) + q_i(k) - \mu \cdot \tilde{F}(k)$.

        end for

    end for

15.4 Simulation

We now give some simulation results, obtained in MATLAB, taken from [122], to illustrate the proposed algorithm.

15.4.1 Consensus and Optimality

According to the previous discussion, assume that the objective is to infer the optimal speed that the ISA system should broadcast to a fleet of EVs traveling in a given area of a city (e.g., in the city center). For this purpose, assume that a fleet of 100 vehicles travels in the city center for an hour, and for the sake of comparison we make the following assumptions:

- In the first 20 minutes, the vehicles travel at the optimal speed calculated from Algorithm 15.1.

- In the second 20 minutes, they travel at a speed above the optimal speed.

- In the last 20 minutes, they travel at a speed below the optimal speed.

In the first stage assume that the communication graph among the EVs changes in a random way, i.e., at each time step an EV receives information from a subset of vehicles belonging to the fleet. This is a simplifying assumption that can be justified by assuming that in principle all vehicles might communicate to all the other vehicles (i.e., they are relatively close), but some communications might fail due to obstacles, shadowing effects, external noise, or some other effects. Besides, in the two last stages we assume that the change of speed occurs almost instantaneously, since there is no requirement to iteratively compute an optimal speed.

The parameters in Algorithm 15.1 as $\eta_i = \mu = 0.001$, and we simulate different cost functions for each EVs by assuming a random number of people inside each car (between 1 and 5 people) with an average weight of 80 kg, and by assuming a different consumption from ancillary services within the typical range of 0.2 to 2.2 kW. The curves of the cost functions used in our experiment are shown in Figure 15.2. Finally, the evolution of the speeds of the EVs are shown in Figure 15.3, while the average energy consumption is shown in Figure 15.4. As can be noticed in Figure 15.4, the optimal speed computed according to Algorithm 15.1 gives rise to the most efficient solution in terms of energy consumption.

15.5   Concluding Remarks

In this chapter we presented an application for speed advisory systems, related to determining the optimal speed that should be followed by a fleet of EVs, with the specific objective of improving their energy efficiency or, in other words, to collaboratively extend their traveling range. The proposed idea has been implemented adopting a distributed consensus algorithm that has the feature to preserve privacy of the personal information, which we believe is an important point to motivate people to effectively collaborate.

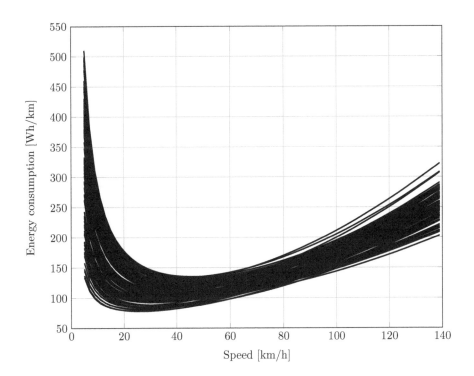

FIGURE 15.2
Utility functions chosen for the set of 100 vehicles. Note that all functions
are convex, and have an individual optimal speed usually between 30 and
40 km/h, which is a reasonable speed for driving in the city center.

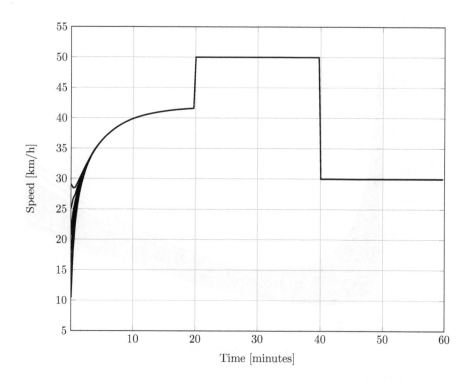

FIGURE 15.3
The vehicles quickly start traveling at the optimal recommended speeds. In
the first 20 minutes the EVs travel at the optimal one.

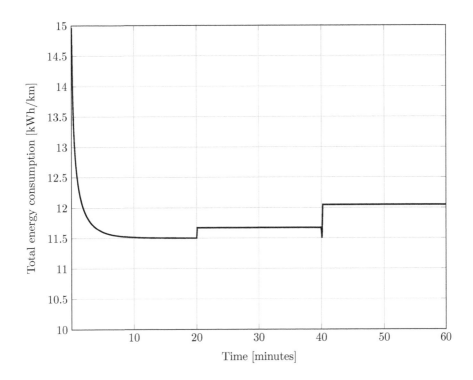

FIGURE 15.4
Energy consumption is minimum when the vehicles travel at the optimal recommended speed

Part IV

Platform Analytics and Tools

16

E-Mobility Tools and Analytics

16.1 Introduction

Most of the problems that we have discussed in the book are characterized by *scale*. In some applications, the benefits of a particular technology are predicated by large scale deployment, and in other applications, our prevailing interest is in developing algorithms to orchestrate the behavior of large scale ensembles of agents. Such issues give rise to new practical and theoretical challenges. Our objective in this part of the book is to briefly touch on these issues, and describe some of the ideas that have been developed and deployed to overcome some of the immediate needs of practitioners.

The first problem that we shall discuss is the need to test large scale deployment of prototype technologies. Clearly, automotive and IoT manufacturers are unable (for reasons of cost) to build thousands of prototype units. On the other hand, large scale simulations alone are unsatisfactory, especially for applications where "human experience" determines eventual acceptability of a particular technology. To bridge the gap between these extremes, a *hardware-in-the-loop* platform has been constructed. This allows system designers to embed real vehicles, driving on real roads, into large scale simulations. In this way it is possible to emulate large scale scenarios, while at the same time giving drivers a *connected car* experience. We give a brief description of this platform, and its potential uses, in Chapter 17.

The final chapter of our book is concerned with bespoke analytics. Many exciting mathematical problems, with exotic constraints and boundary conditions, arise in an ITS context. Examples of applications include the need to predict driver intent, the need to guarantee ergodic behavior of large scale ensembles of agents, the need to solve large scale optimization problems in a distributed manner, while at the same time coping with variations in the number of agents, coping with closed loop data sets in real time, preserving the privacy of individual agents, and not over-burdening the communication network. The final chapter gives some hints as to design feedback systems to solve certain optimization problems in a scale free manner; that is, when the number of agents participating in the scheme is unknown. In the first of these approaches these objectives are achieved using only intermittent feedback.

Notes and References

Here, we note our debt of gratitude to the following co-workers who worked on the topics described in this part of the book. The chapter on the hardware-in-the-loop platform is based on paper [77], in collaboration with Wynita Griggs, Rodrigo Ordóñez-Hurtado, Florian Häusler and Kay Massow[1]. The chapter on AIMD is based on work with Jia Yuan Yu and Martin Corless. The interpretation of passivity is based on joint work with Brian Anderson and Wynita Griggs.

---

[1]©IEEE. Reprinted, with permission, from [77].

17

A Large-Scale SUMO-Based Emulation Platform

17.1 Introduction

Mobility simulators are an essential tool to design and evaluate new applications and services in the context of ITS [156]. Our objective here is to use one such tool, SUMO, in conjunction with real vehicles, to evaluate applications that we have developed and described throughout the book. SUMO [113] is an open source, microscopic road traffic simulation package primarily being developed at the Institute of Transportation Systems at the German Aerospace Centre (DLR). SUMO is designed to handle large road networks, and comes with a "remote control" interface, TraCI [189], that allows one to adapt the simulation and to control singular vehicles on the fly. SUMO allows for different ways to generate road networks: they can be defined by the user in XML; abstract road networks can be generated using the SUMO application *netgenerate*; or they can be imported from different formats (e.g., from OpenStreetMap). Similarly, vehicular flows can be (manually) given by the users (e.g., by providing an origin to destination matrix, or by fixing junction turning probabilities); alternatively, more realistic historical data from some sample cities have been made available by researchers in the ITS community, and can be uploaded in the SUMO environment.

While mobility simulators can be used to simulate large scale road networks, they cannot accommodate for all the complexities, uncertainties, technical issues, and drivers' attitudes and responses that might arise in the real world [183]. For this reason, proof of concepts prototypes and vehicles are also used to further demonstrate ITS applications. However, large fleets of (thousands of) vehicles equipped with the prototype technologies and communication abilities necessary for testing (e.g., connectivity-based) ITS applications are obviously impractical. At the same time, while real-world test fleets of small numbers of vehicles may be enough to demonstrate some elements of proof of concept applications, they are not suited for many applications requiring much larger fleet size and city-wide scenarios. One way to address this problem is to merge both real vehicles and simulation. In this chapter we describe a prototype simulation-based platform that we have built for embedding, in real time, real vehicles into SUMO. The objective is to provide the drivers of real cars with a real feeling of being in a connected car use case, and thus, to somewhat experience how it feels to be part of a large-scale connected vehicle scenario. We call this platform a Hardware-in-the-Loop (HIL) platform. Roughly speaking, the HIL platform is constructed by embedding real vehicles (traveling on real roads) in a simulation platform. Measurements from both the real vehicles and the virtual ones are used to advise the vehicle fleet for a given use case. The HIL platform is designed with the following objectives in mind. First, the HIL simulation platform should permit real human driver response (to the advice and directions given by the simulator), to be observed and fed back to the simulator in real time, so that the simulator may process human inputs that cannot otherwise be possibly predicted. Second, the HIL should permit human behavioral factors to be explored in a safe environment. Third, the HIL platform should incorporate elements of vehicle safety into testing with real drivers (for example, for non line-of-sight incidents). Finally, provided that a real vehicle is available, the platform should be inexpensive to construct.

The remainder of this chapter is organized as follows. In Section 17.2, we review previous work on this topic. In Section 17.3, we describe the key components of the HIL platform. In Section 17.4, we briefly describe a sample application that has been implemented using the platform. Finally, possible interesting further lines of research are identified in Section 17.5.

17.2   Prior work

Three main categories of simulators can be discerned in the literature: traffic simulators, driving simulators, and networking simulators. Traffic simulators are typically used to develop strategies to improve the mobility and safety of urban and rural travel; driving simulators are convenient to test human responses to new applications in a controlled environment; and finally, networking simulators are used to evaluate the ability of Vehicular Ad-hoc NETworks (VANETs) to exchange information in a challenging environment, like the urban one, in order to implement new cooperative strategies (e.g., cooperative routing, safety alerts, and congestion messages). In this latter context, recent years have witnessed a growing interest in developing more realistic environments to test and validate ITS applications. For example, [149] illustrates preliminary benefits in integrating driving and traffic simulators. Clearly, mobility simulators would benefit from realistically taking into account drivers' behaviors, which is considered in driving simulators, while driving simulators would provide an improved experience if implemented in a reasonably believable traffic environment. Similarly, [53] describes efforts aiming to develop a virtual IntelliDrive testbed within a microscopic traffic simulator. IntelliDrive is a kind of V2I integrated platform that uses advanced wireless technologies to implement applications such as "systems warning drivers of traffic slowdowns ahead, systems warning about cross-street vehicles that may potentially run through a red light, and systems notifying drivers of roadway features, such as sharp curves". The authors in [148] describe the modeling and the implementation of an architecture to integrate a robotics and a traffic simulator to facilitate the study of the driving behavior of autonomous vehicles in human-steered urban traffic scenarios. The integration of more simulators was also investigated in [128], where the Simulink/MATLAB model of an electric bus powertrain subsystem was exported into a mobility simulator. Finally, similar concepts are further extended in [204] where traffic, driving and networking simulators are all integrated into one only research and development tool, which is used to implement new applications for connected vehicles[1].

Other works in the literature have attempted to increase the level of realism of current simulators by considering some real vehicles traveling in the road network as well. In this context, we note [152]. In this work, the objective was to implement a mixed reality platform, where intersection control policies for autonomous vehicles (formerly only tested in simulation) were tested with a real autonomous vehicle, interacting with multiple virtual vehicles (in a simulation), at a real intersection, in real time. In such an example, having all real vehicles would have been expensive in case some control policies failed. At the same time, the experiment proved to provide results that were different from those obtained using a fully simulated environment. Similarly, [131] describes GrooveNet, which is a hybrid V2V network simulator, capable of communication between simulated vehicles, real vehicles, and between real and simulated vehicles. With such an approach it becomes feasible to deploy a small fleet of vehicles (in the example, in the order of a dozen), to test protocols that in truth involve hundreds or thousands of vehicles, which are simulated. The HIL approach of [131] is close to what we describe here, though GrooveNet is designed with other objectives in mind. Here, we are interested in providing a platform in which ITS applications can be investigated and experienced by a human driver. In particular, we are most interested in investigating closed-loop applications (involving feedback). On the other hand, GrooveNet was designed to investigate V2V issues; in particular, with respect to wireless communication issues in mobility networks. This latter issue is of no interest in our context, while the applications of our interest, which are in essence closed-loop control applications, can be investigated with a greater level of ITS realism using the SUMO-based platform.

---

[1]see for example VSimRTI - Smart Mobility Simulation - available online at https://www.dcaiti.tu-berlin.de/research/simulation. Last Accessed July 2017.

17.3   Description of the Platform

We now describe the main components of the HIL platform. A general overview of the architecture of the platform is shown in Figure 17.1. The main objective of the platform architecture is to provide the possibility for feedback between simulated vehicles and the human driver in the real vehicle. The main components are the applications deployed on the base station computer interfacing hundreds of vehicles in SUMO, and the SumoEmbed component, deployed on the smartphone, which serves as the interface to the driver and the vehicle.

Real Vehicle

The original field-test vehicle used in the platform was a 2008 Toyota Prius 1.5 5DR Hybrid Synergy Drive and is pictured in Figure 17.2. More recently a 2015 Plug-in Toyota Prius has been used, but any vehicle with an accessible interface or gateway (such as an OBD-II diagnostic connector, in our case) is suitable to be used as a field-test vehicle. The specifics of the ITS application being tested may place further restrictions on the vehicle choice. The road network we generated in SUMO. A virtual vehicle representing the Toyota Prius (i.e., an avatar of the real vehicle) was created in SUMO. This was partially achieved by assigning physical characteristics to the virtual vehicle that were approximately the same as those of the real car. Further details on defining vehicle types and routes in SUMO are available in the user documentation found on the SUMO website[2].

Road Network

For the applications tested in [77] road networks were imported from OpenStreetMap. In particular, the maps using Java OpenStreetMap Editor (JOSM)[3] were cleaned with XMLStarlet[4] before applying SUMO's *netconvert*.

Smartphone

In the real vehicle, a Samsung Galaxy S III mini (model no. GT-I8190N) running as operating system Android Jelly Bean (version 4.1.2) was used as an interface to the vehicle. The purpose of the smartphone is to relay, over a cellular network, periodic information from the vehicle's onboard computer (e.g., the speed of the vehicle) to the base station computer running SUMO, and to receive messages from the base station computer and display them on the smartphone user interface for the driver (e.g., recommendations for alter-

---

[2]`http://www.sumo-sim.org`. Last Accessed July 2017.

[3]JOSM (Java OpenStreetMap Editor). Website: `http://josm.openstreetmap.de`. Last Accessed July 2017.

[4]Website: `http://xmlstar.sourceforge.net`. Last Accessed July 2017.

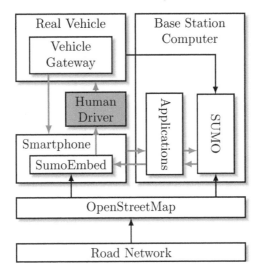

FIGURE 17.1
Main components of the platform. The gray arrows highlight the dynamic connections (i.e., the feedback loop) between the components.

native driving behaviors). The mobile data services of a commercial mobile phone operator are used, where these services were provided using a 3G UMTS 900/2100 network, for the relay of data. The plug-in application that we developed for the smartphone is called SumoEmbed and is described further below.

Python Script

Application-specific scripts in Python 2.7.3 that were run simultaneously with SUMO on the base station computer, are used to interface SUMO and the vehicle. The scripts typically consist of two parts: (i) a main part that acts as a client to SUMO and adapts the traffic scenario simulations, according to the ITS application being implemented, online via TraCI (TraCI uses a TCP-based client/server architecture to provide access to SUMO); and (ii) a second part that acts as a TCP server, listening for incoming calls from the smartphone and then handling the data transfer between the smartphone and the base station computer running SUMO.

Vehicle Gateway

The hardware device used to connect the smartphone and the Toyota Prius' onboard computer was the Kiwi Bluetooth OBD-II Adapter by PLX Devices[5]

---

[5]PLX Devices Inc., 440 Oakmead Parkway, Sunnyvale, CA 94085, USA. Phone: +1 (408) 7457591. Website: http://www.plxdevices.com. Last Accessed July 2017.

FIGURE 17.2
Field-test vehicle: 2008 Toyota Prius

The device plugs into the vehicle's OBD-II diagnostic connector and communicates to the smartphone via Bluetooth. A variety of existing smartphone applications are compatible for use with Kiwi Bluetooth. Among those Torque Pro[6] was chosen given that an Android Interface Definition Language (AIDL) Application Programming Interface (API) was included to handle third party plug-in applications.

Before proceeding, some comments are appropriate, as described in the following subsection.

Remark 17.1. In the prototype design, a single real vehicle was embedded into SUMO. To do this in practice, we exploited the fact that in the parent process of our Python script, TraCI provides access to SUMO via a single port, and multiple real vehicles can be represented in SUMO by different vehicle IDs. Passing or sharing the information of multiple vehicles between this parent process of the Python script, and our Python subprocess that communicates with the smartphone, may require the use of, for example, advanced shared memory formats in Python. A single port was reserved for communication with the smartphone in our Python subprocess, but it would be very simple to extend this to handle multiple incoming calls (i.e., from multiple smartphones). The specific information exchanged between the real vehicle and the simulator is application-driven and, in our current set-up, depends on what data can be sent via the vehicle gateway (e.g., OBD-II diagnostic connector) and/or via the smartphone (e.g., human input through a user interface), and whether this information needs to be concurrently managed in SUMO. In principle, any information on the OBD-II can be used in the emulation set-up. Information between the real vehicle and the base station computer, in our current set-up, is transmitted once every second. This steady, periodic rate is

---

[6] *Torque Pro* by Ian Hawkins. Available from Google Play: https://play.google.com/store/apps/details?id=org.prowl.torque. Last Accessed July 2017.

maintained throughout the duration of the simulation, for simplicity. However, the information transmission frequency can be changed to implement other applications as well, e.g., to enable event-driven information transmission.

17.4 Sample Application

The HIL platform has been used to test a range of ITS applications, some of which have been described in early chapters (e.g., Chapters 4, 13 and 14). As a further example, we now, very briefly, revisit the speed advisory problem described in Chapter 15 with full details of the application described in [122]. In that paper, such a system was designed to minimize emissions along a given route for regular ICEVs. To validate the design, the authors had used the HIL platform outlined in this chapter. Specifically the Prius test vehicle was embedded into a network of 30 ICEVs (i.e., with 29 other test ICEVs). The emissions class for each of the virtual vehicles was selected randomly from a set of profiles defined in [20]. The objective of the HIL tests was to record the ability and willingness of the Prius driver to follow the recommended speed. In the example, the ISA system starts recommending an optimal speed after 300 seconds of motion. A trace depicting the driver behavior in response to such recommendation is shown in Figure 17.3, while in Figure 17.4 the corresponding reduction of CO_2 emissions is shown.

17.5 Concluding Remarks

The HIL platform for emulating large-scale intelligent transportation systems was presented. The platform embeds a real vehicle into SUMO. A goal of the platform is to provide drivers with some sense of how it would feel to participate in large-scale, feedback-based, connected vehicle applications, and thus allow ITS developers to better examine real driver reactions in regards to feedback control for applications in an ITS context.

The platform illustrated here can be generalized and extended along a number of lines. In particular, interesting enhancements would be (i) a better map-matching for those applications that are location-aware; (ii) the integration of the platform with a discrete-event network simulator (e.g., ns-3), the inclusion of V2V communication capabilities, and the addition of further embedded real vehicles; (iii) improvements in the modular design of the platform; and (iv) an enhancement of the platform with the ability to incorporate real-time traffic information. Also, it would be interesting to explore other ITS applications that might benefit of the main advantage of our platform, i.e., the

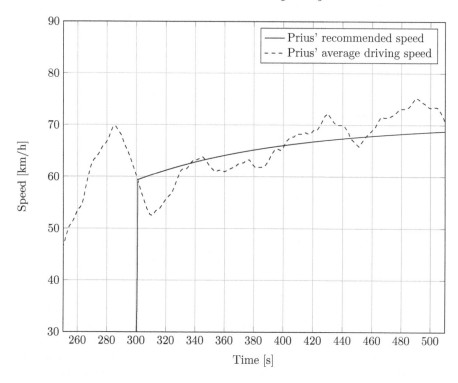

FIGURE 17.3
Evolution of the speed related with the Prius for the HIL simulation. The
algorithm was turned on around time 300 s. The average speed was calculated
with a window size of 20 time steps.

ability to investigate the feedback-loop relationship between the environment
and the vehicle.

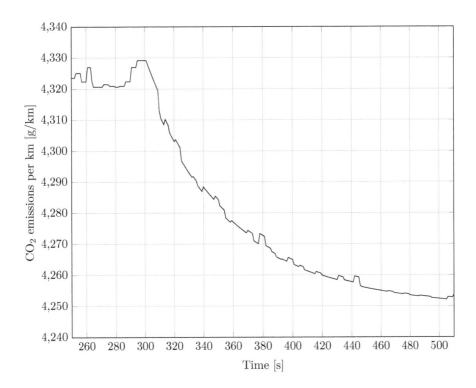

FIGURE 17.4
Evolution of reduction of overall CO_2 emissions. The algorithm was turned on around time 300 s.

18

Scale-Free Distributed Optimization Tools for Smart City Applications

18.1 Introduction

In almost all applications investigated in this book, the problem arises of provisioning a limited resource in a distributed manner, without explicit knowledge of the number of agents attempting to share that resource. Moreover the number of agents can grow very large. Such problems can be formulated as optimization problems or as feedback control problems, depending on the application. One issue in the design of such systems is the need to find control algorithms whose actions do not depend on the (unknown) number of agents participating in the scheme. This independence of the number of agents assures the scalability of the control algorithm. Further related issues that the control algorithm has to handle are privacy concerns and communication overhead. In this chapter we shall describe two methodologies that can be employed in such circumstances.

18.2 The AIMD Algorithm

Consider n agents which all aim to access the same resource. The precise quantity of available resource is unknown to the agents and it is not practical for the agents to communicate with each other. The additive-increase multiplicative-decrease (AIMD) algorithm has been deployed with some success in such situations.

In the continuous-time version of the AIMD algorithm each agent i has an internal variable $x_i(t)$ which at time t represents the quantity of shared resource used by agent i. Two parameters, $\alpha_i > 0, \beta_i \in [0,1)$, characterize the time behavior of agent i. The algorithm consists of two alternating phases. In the additive increase phase the consumption is increased linearly in time with slope α_i, so that if $x_i(t_k)$ was the consumption at the beginning of an additive increase phase, then

$$x_i(t) = x_i(t_k) + \alpha_i(t - t_k) \tag{18.1}$$

is the consumption at time $t > t_k$ in the same additive increase phase. As all agents continuously increase their consumption of the limited resource this limit will eventually be reached. We call the occurrence of this a capacity event and we label the times of capacity events by $t_1 < t_2 < \ldots < t_k < \ldots$. At capacity events agents have to be made aware of such an event; the precise nature of this communication has to depend on a particular application. Those agents that are aware will decrease their consumption instantaneously in a multiplicative fashion so that

$$x_i(t_{k+1}^+) = \beta_i x_i(t_k) = \beta_i \lim_{t \nearrow t_{k+1}} x_i(t). \tag{18.2}$$

Combining (18.1) and (18.2) we see that the evolution of agent i from capacity event t_k to capacity event t_{k+1} is modeled by

$$x_i(t_{k+1}) = \beta_i x_i(t_k) + \alpha_i(t_{k+1} - t_k). \tag{18.3}$$

This relation can be extended to a full model of the evolution of all agents by eliminating the time between capacity events $t_{k+1} - t_k$. Assuming that the amount of available resource is given by a constant C_r we have at all capacity events t_k that

$$C_r = \sum_{j=1}^{n} x_j(t_k) = \sum_{j=1}^{n} x_j(t_{k+1}) = \sum_{j=1}^{n} \left(\beta_j x_j(t_k) + \alpha_j(t_{k+1} - t_k) \right). \tag{18.4}$$

Thus

$$t_{k+1} - t_k = \frac{1}{\sum_{j=1}^{n} \alpha_j} \sum_{j=1}^{n} (1 - \beta_j) x_j(t_k). \tag{18.5}$$

And inserting this into (18.3) we obtain

$$x_i(t_{k+1}) = \beta_i x_i(t_k) + \frac{\alpha_i}{\sum_{j=1}^{n} \alpha_j} \sum_{j=1}^{n} (1 - \beta_j) x_j(t_k). \tag{18.6}$$

It is now convenient to introduce the vector $x(t) \in \mathbb{R}^n$ with entries $x_i(t)$ and the matrix

$$\mathbf{A} := \begin{bmatrix} \beta_1 & & 0 \\ & \ddots & \\ 0 & & \beta_n \end{bmatrix} + \frac{1}{\sum_{j=1}^{n} \alpha_j} \begin{bmatrix} \alpha_1 \\ \vdots \\ \alpha_n \end{bmatrix} \begin{bmatrix} (1 - \beta_1) & \cdots & (1 - \beta_n) \end{bmatrix}. \tag{18.7}$$

With this notation (18.6) for $i = 1, \ldots, n$ can be summarized as

$$x(t_{k+1}) = \mathbf{A} x(t_k). \tag{18.8}$$

At first it is surprising that the AIMD algorithm results in such a simple linear iteration, at least in the synchronized case in which all agents react to all capacity events. All of the entries of $\mathbf{A}$ are positive and the columns of $\mathbf{A}$ sum to 1, so that the matrix is column stochastic. The Perron-Frobenius theorem can now be used to deduce several important features of the iteration (18.8). These properties and further investigations are derived in full detail in the first chapters of [39].

Remark 18.1. The above analysis assumes that all agents respond in unison to a capacity notification. In reality, only some agents will do this, and usually the response of an agent to a capacity notification is characterized in a probabilistic manner. Such situations are also described by a linear system; this time a stochastic switched linear system. Details of this unsynchronized model can be found in [39, Chapters 6-9].

18.3   Optimal Resource Allocation

The AIMD algorithm can also be used to solve certain optimization problems. Let us assume again that n agents wish to share a constant resource of which $C_r > 0$ units are available. Each agent associates a cost $f_i(x_i)$ to the use of a quantity $x_i \geq 0$ of the resource. Alternatively, agents can associate a utility $g_i(x_i)$ to using a certain amount of resource. There is no substantial difference in studying these problems: one would like to minimize costs and on the other hand maximize utility. The negative of a cost function thus becomes a utility function and vice versa.

An optimal resource allocation problem can be formulated as

$$\begin{cases} \min_{x_1,\dots,x_n} & \sum_{i=1}^{n} f_i(x_i) \\ \\ \text{subject to} & \sum_{i=1}^{n} x_i = C_r, \quad \text{and} \quad x_i \geq 0, \quad i = 1,\dots,n. \end{cases} \tag{18.9}$$

For $i = 1,\dots,n$ we assume that the cost functions $f_i : [0, C_r] \to \mathbb{R}$ are strictly convex and continuously differentiable that satisfy $f_i(0) = 0$. The constraints of this optimization problem express two major properties we have in mind: (i) the resource consumption x_i can only be a nonnegative value, so agents cannot produce the resource, (ii) the consumption should use all the available resource C_r so that the allocation is not wasteful, (iii) there is a hard upper bound on the available resource. In the earlier chapters of this book, examples of the cost functions have been the inconvenience (or the price) for the owners of EVs to charge their vehicles at given times in Chapter 7; or the pollution emitted by a given ICE vehicle in Chapter 13; or the energy consumption of an EV in Chapter 14. In some cases, we have also been interested in solving maximization problems where the (now concave) functions could be interpreted as utility functions (e.g., savings of CO_2 emissions). Note that in all the previous examples, our x_i variables were physical (non-negative) quantities, e.g., money, energy, or emissions, and this gives rise to the non-negative constraint in (18.9).

It is well known that compactness of the feasible space implies that an optimal point x^* at which the minimum is attained exists, and that the optimal point is unique as a consequence of the assumption of strict convexity. In particular, the unique point is known to satisfy the Karush-Kuhn-Tucker (KKT) conditions over the Lagrangian equation associated with (18.9), see for instance [22]. However, here we are not interested in the conventional centralized solution of problem (18.9), but rather, we shall show that AIMD can be used to obtain convergence of the long-term average state to the KTT point x^*. In particular, convergence to the optimal solution can be achieved in a distributed fashion, requiring little communication of relevant data, and in a privacy-preserving manner.

This objective can be achieved by introducing a stochastic aspect in the algorithm. We introduce variable probabilities $p_i, i = 1, \ldots, n$ which define for each agent the probability of reducing their state if a capacity event is announced. The key is to customize these probabilities p_i of single agents based on the past behavior of that agent and its cost function f_i. More specifically, we let $x_i(k)$ denote the value of the state at the kth capacity event, and $\overline{x}_i(k)$ denote the long-term average of the state variables, i.e.,

$$\overline{x}_i(k) = \frac{1}{k+1} \sum_{j=0}^{k} x_i(j). \tag{18.10}$$

We then require that each agent performs the decrease step with probability

$$p_i\left(\overline{x}_i(k)\right) = \Gamma \frac{f_i'\left(\overline{x}_i(k)\right)}{\overline{x}_i(k)}. \tag{18.11}$$

In equation (18.11) $f_i'(\cdot)$ denotes the derivative of the cost function, and Γ is a suitable constant ensuring that $0 \leq p_i(x_i) \leq 1$ for all $x_i \in [0, C_r]$ and $i = 1, \ldots, n$. This of course imposes further conditions on the formulation of the optimization problem. The f_i' should be positive, strictly increasing and bounded on the interval $[0, C_r]$. Also such a bound needs to be known so that a suitable constant Γ can be chosen. This however imposes only small constraints on the class of optimization problems, see [193] for a discussion of reformulations of a given optimization problem so that the condition in (18.11) can be applied.

Equation (18.11) implies that each user will respond to the notification of a capacity event independently of the behavior of the other agents with a customized probability $p_i\left(\overline{x}_i(k)\right)$. The response is then to perform the multiplicative decrease step on its share of the resource. It can be shown that the following convergence to the optimum holds (almost surely, that is, with probability one) provided that all agents use the same parameters α, β in their implementation of the AIMD algorithm (see [193] for detailed proofs in this regard):

$$\lim_{k \to \infty} \overline{x}(k) = x^*.$$

In particular, note that the result guarantees that both the limit of $\overline{x}_i(k)$ exists, and that in fact it is equal to the optimal value. Underlying this result is the observation that AIMD algorithms attempt to achieve *consensus* among participating agents. Now for the particular problem under consideration here the KKT conditions which characterize the optimal point are

$$f_i'(x_i^*) = f_j'(x_j^*), \quad \text{for all indices } i, j = 1, ..., n.$$

So in the optimal point the derivatives of the cost functions are indeed in consensus and x^* is the only feasible point with this property. This condition follows from the KKT conditions, and is consistent with the optimality of the steady-state vector of states obtained with the AIMD algorithm.

Remark 18.2. Note that in some cases the AIMD algorithm, and more specifically Equation (18.11), may be reformulated for convenience by shifting the derivative terms $f_i'(\cdot)$ to the denominator.

18.4   Scale-Free Advantages of AIMD

In principle, many different methods may be used to solve the optimization problem (18.9) to compute the optimal way to share the available resource among the interested agents. For instance, in Chapter 13 we have deployed a centralized PI-like solution to share a "budget" of pollution (resource) among the traveling vehicles (agents). In other cases (see for instance Chapter 14) we have alternatively used AIMD. The choice of AIMD is useful for many applications due to its low-communication requirements, privacy-preservation properties and the fact that AIMD requires very little actuation ability on the agent-side. This is in contrast to other distributed algorithms, like the PI-like control or ADMM (see [21], where at each time step, agents must solve a local optimization problem). However, the main advantage of AIMD appears to be the following:

Algorithm parameterization: *In AIMD the gain parameters of the network are independent of network dimension; rather, they only depend on the largest derivative over all utility functions* (i.e., the parameter Γ in (18.11)). Thus, selecting a gain for the algorithm is extremely simple in the case of AIMD.

A pseudo-code to implement the AIMD algorithm for our applications in general is now briefly reported: Note that the algorithm is performed in an

---

Algorithm 18.1 Unsynchronized AIMD Algorithm

1: **Initialization:** $k = 1$, $x_i(k) = 0$;
2: Broadcast the parameter Γ to the entire networks;
3: **while** $k < k_{\max}$ **do**
4:     **if** $\sum_{i=1}^{N} x_i(k) < C_r$ **then**
5:         $x_i(k+1) = x_i(k) + \alpha$
6:         **else** with probability $p_i(k) = \Gamma \dfrac{f_i'\left(\overline{x}_i(k)\right)}{\overline{x}_i(k)}$, $x_i(k+1) = \beta x_i(k)$
7:         **else** $x_i(k+1) = x_i(k) + \alpha$
8:     **end if**
9:     $k = k+1$
10: **end while**

---

iterative way, and here $k_{\max}$ represents the maximum number of iterations before the algorithm stops. According to the AIMD spirit, if the sum of shares $x_i(k)$ of the shared resource is smaller than C_r, then each agent increases its share by a quantity α. However, if the sum of the shares exceeds C_r (i.e., capacity event), then an agent decreases its share by a multiplicative factor $0 < \beta < 1$ with probability $p_i(k) = \Gamma\frac{f_i'(\overline{x}_i(k))}{\overline{x}_i(k)}$, or keeps increasing its share with probability $1 - p_i(k)$. It is then proved in [193] that $\overline{x_i}(k)$ approaches the optimal solution of the optimization problem when Algorithm 18.1 converges.

Remark 18.3. We note that AIMD has some nice properties concerning the privacy of the agents. First, there is no inter-agent communication and only limited feedback is required. Second, the algorithm is implicit, i.e., only requiring a consensus on the utility functions. Finally, it is stochastic, so making reconstruction of the f_i's is more difficult from observations. See [137] for a more detailed discussion.

18.5   Passivity

Another scale free approach from the control literature comes from the notion of passivity. A passive linear time-invariant system is a system whose transfer function is positive real. The notion of a positive real (PR) function and of a strict positive real (SPR) function is important in many areas of engineering systems, in particular in control theory and in circuit theory. In circuit theory the term passive component, i.e. an element that consumes but does not supply energy, is commonly used. This use illustrates the relation between passivity and energy which is often exploited in physical systems. An important historical property that has made passivity (and positive realness) particularly attractive to classical control engineers is that a physical system, which is passive, has properties that makes the behavior of the system "friendly". For example, it is well known that a negative feedback connection of two (strictly) passive systems is always asymptotically stable, and that the stability of more complicated interconnections of passive systems is characterized by simple algebraic conditions [17, 136]. This has made passivity very useful in the design of distributed control systems. More recently, passivity has assumed an important role in the study of optimization algorithms and consensus [190], the study of cyber-physical systems [70], and in the exploration of diagonal stability problems [17].

Our objective in this brief section is to make readers aware that passivity may be used as a design tool for designing closed loop systems. There are many ways of understanding passivity. One can talk about generalized notions of energy, and one may also speak of ensuring that in a negative feedback loop a phase shift of less than π radians is always satisfied. Perhaps the simplest method of explaining passivity is however to appeal to Lyapunov arguments and it is precisely these arguments that we shall describe in this book. To this end we recall the following from standard linear systems theory. In what follows we assume that readers are familiar with elementary systems theory: see for example, the textbook [104] for a revision of some of these concepts.

Consider a linear time-invariant input-output system:

$$\begin{aligned} \dot{x} &= \mathbf{A}x + \mathbf{B}u \\ y &= \mathbf{C}x + \mathbf{D}u \end{aligned} \tag{18.12}$$

where $\mathbf{A} \in \mathbb{R}^{n \times n}$ and $\mathbf{B}, \mathbf{C}, \mathbf{D}$ are matrices of dimension $n \times m$, $m \times n$ and $m \times m$ respectively. We assume that the above system is stable; that is that the matrix $\mathbf{A}$ has eigenvalues in the open left half of the complex plane. Associated with this linear system is its transfer function matrix:

$$\mathbf{G}(s) = \mathbf{C}(s\mathbf{I} - \mathbf{A})^{-1}\mathbf{B} + \mathbf{D}. \tag{18.13}$$

Historically, most classical criteria for stability had been established in terms of the frequency response of this transfer function matrix $\mathbf{G}(j\omega)$, $-\infty < \omega < \infty$ and j is the imaginary unit. For example, a system is said to be strictly passive if $\mathbf{G}(j\omega) + \mathbf{G}(j\omega)^* > 0$ for all ω, and the classical Nyquist criteria states that the feedback interconnection of two such Linear Time Invariant (LTI) systems $\mathbf{G}_1$ and $\mathbf{G}_2$ is stable if and only if [205] the Nyquist plot of $\det[\mathbf{I} + \mathbf{G}_1(j\omega)\mathbf{G}_2(j\omega)]$ does not make any encirclements of the origin.

Suppose now that $\mathbf{G}_1(j\omega)$ and $\mathbf{G}_2(j\omega)$ are both passive. Then, for all $\omega \in \mathbb{R}$

$$\begin{aligned} \mathbf{G}_1(j\omega) + \mathbf{G}_1(j\omega)^* &> 0 \\ \mathbf{G}_2^{-1}(j\omega) + \mathbf{G}_2^{-1}(j\omega)^* &> 0. \end{aligned} \tag{18.14}$$

(where invertibility follows directly from the Lyapunov equation). Thus:

$$\mathbf{G}_1(j\omega) + k\mathbf{G}_2^{-1}(j\omega) + \mathbf{G}_1(j\omega)^* + k\mathbf{G}_2^{-1}(j\omega)^* > 0 \qquad (18.15)$$

for all $k \geq 0$ (and in particular for $k \geq 1$). For each fixed ω, Lyapunov's inertia theorem implies that $\mathbf{G}_1(j\omega) + k\mathbf{G}_2^{-1}(j\omega)$ cannot have eigenvalues on the imaginary axis. Thus,

$$\det\left[\mathbf{G}_1(j\omega)\mathbf{G}_2(j\omega) + kI\right] \neq 0. \qquad (18.16)$$

Since this is true for all $k \geq 1$, and since $\mathbf{A}_1, \mathbf{B}_1, \mathbf{C}_1, \mathbf{D}_1$, $\mathbf{A}_2, \mathbf{B}_2, \mathbf{C}_2, \mathbf{D}_2$ (i.e., the matrices corresponding to the state-space representation of transfer function matrices $\mathbf{G}_1$ and $\mathbf{G}_2$) are real matrices, then this implies that $\det[\mathbf{I} + \mathbf{G}_1(j\omega)\mathbf{G}_2(j\omega)]$ does not make any encirclements of the origin (by continuity). Thus, strict passivity is a sufficient condition for Nyquist to hold.

In addition, consider the case where $\mathbf{G}_1(j\omega)$ is a control and $\mathbf{G}_2(j\omega)$ is a parallel connection of passive elements. In this case $\mathbf{G}_2(j\omega)$ is the sum of passive elements, which is also a passive element. Thus, adding or removing one or more elements will not affect the stability of the feedback loop.

In particular, note that all the previous statements guarantee that appropriate interconnections of passive LTI systems form an overall stable system, independently of the size of the single LTI subsystems.

18.6   Concluding Remarks

In most of the applications investigated in this book, the problem arises of provisioning a limited resource in a distributed manner, without explicit knowledge of the (possibly very large and time-varying) number of agents attempting to share that resource. As we have shown, these problems can be formulated as optimization problems, o feedback control problems in general. However, a challenging aspect regards the design of optimal control algorithms whose actions should not depend on the (possibly unknown) number of agents participating in the scheme. Achieving this size-free property, together with handling privacy concerns and communication overhead, motivated the discussion of the two methodologies, namely, AIMD and passivity, provided in this chapter.

Postface

As we mentioned in the Preface, this book describes some of the work carried out by the authors and their co-authors in the period 2011-2017. Our objective in writing this book, and in much of our other work, was not to focus on the details of EV propulsion mechanisms (this is covered in detail elsewhere), but rather to focus on the opportunities of the new networked actuation possibilities afforded by FEVs and PHEVs in our cities. While we have included many use cases to illustrate this point, others, which we believe to be important, are omitted due to space limitations, such as [91, 81]. In particular, the migration of ideas to PEDELECS, and the focus on protecting cyclists, would appear to be a very fruitful and useful direction for future research; see [178] for some initial ideas in this direction.

One of the main reasons for our original interest in this topic was the role that EVs could have in addressing a spectrum of societal challenges. Looking back, our expectations in 2011 promised rapid adoption of the green agenda, and equally rapid adoption of EVs in towns and cities, all as part of a bigger Smart City and Smart Grid revolution. In this context, as we write this last piece of text (July 2017), it is very hard not to be disappointed. Except in a few outlier countries, EV adoption has been slow, and the transportation agenda is still driven by greenhouse gas concerns (and only to a much lesser extent by air quality related issues). Notwithstanding this fact, we believe that there are grounds for optimism. Strong companies such as TESLA and FARADAY FUTURE have emerged, conventional OEMs are now well on the path to fleet electrification, and the issue of *air quality*, which is perhaps the defining issue of our age, is now becoming a topic of great concern worldwide. In addition, very recent announcements from car companies[1] and national governments[2,3] seem to confirm that E-mobility is eventually gaining momentum. As we place our work in this context, we hope that this book will find its natural place, and perhaps help practitioners and theoreticians push back some of the new barriers to EV adoption.

---

[1]https://www.theguardian.com/business/2017/jul/05/volvo-cars-electric-hybrid-2019. Last Accessed July 2017.

[2]https://www.theguardian.com/business/2017/jul/06/france-ban-petrol-diesel-cars-2040-emmanuel-macron-volvo. Last Accessed July 2017.

[3]http://www.telegraph.co.uk/news/2017/07/25/new-diesel-petrol-cars-banned-uk-roads-2040-government-unveils/. Last Accessed July 2017.

Finally, it is worth noting that some, if not all, of the work here, informed and motivated some exciting theoretical research directions. Dimensioning issues for sharing systems, privacy-preserving optimizations, scale-free feedback control, are all issues that we have grappled with during the course of our work. Pressing issues such as the need for ergodic feedback design, the co-design of prediction and feedback systems, the need for closed loop identification at scale, and the need to predict driver intent in a closed loop, are all important themes that remain to be resolved [65, 44] It is with this latter thought in mind that we look forward with great excitement to the next part of our E-mobility journey.

Emanuele Crisostomi
Pisa, July 2017

Robert Shorten
Dublin and Berlin, July 2017

Sonja Stüdli
Newcastle (Australia), July 2017

Fabian Wirth
Passau, July 2017

References

[1] S. Acha, T.C. Green, and N. Shah. Effects of optimised plug-in hybrid vehicle charging strategies on electric distribution network losses. In *Proc. IEEE PES Transmission and Distribution Conference and Exposition*, pages 1–6, New Orleans, LA, USA, April 2010.

[2] S. Acha, T.C. Green, and N. Shah. Optimal charging strategies of electric vehicles in the UK power market. In *Proc. IEE PES Innovative Smart Grid Technologies (ISGT)*, pages 1–8, Anaheim, CA, USA, January 2011.

[3] E. Adell, A. Várhelyi, M. Alonso, and J. Plaza. Developing human-machine interaction components for a driver assistance system for safe speed and safe distance. *IET Intelligent Transport Systems*, 2(1):1–14, 2008.

[4] I. Adler and R.D. Monteiro. A geometric view of parametric linear programming. *Algorithmica*, 8(1):161–176, 1992.

[5] A. Afroditi, M. Boile, S. Theofanis, E. Sdoukopoulos, and D. Margaritis. Electric vehicle routing problem with industry constraints: trends and insights for future research. *Transportation Research Procedia*, 3:452–459, 2014.

[6] European Environment Agency. Monitoring CO_2 emissions from new passenger cars in the EU: Summary of data for 2012. European Environment Agency, Copenhagen, Denmark, 2013. Available online at http://www.eea.europa.eu/publications/monitoring-co2-emissions-from-new-cars, last accessed: July 2017.

[7] A. Ahlbom, J. Bridges, R. De Seze, L. Hillert, J. Juutilainen, M.-O. Mattsson, G. Neubauer, J. Schüz, M. Simko, and K. Bromen. Possible effects of electromagnetic fields (EMF) on human health–opinion of the scientific committee on emerging and newly identified health risks (SCENIHR). *Toxicology*, 246(2–3):248–250, 2008.

[8] F. Alesiani and N. Maslekar. Optimization of charging stops for fleet of electric vehicles: a genetic approach. *IEEE Intelligent Transportation Systems Magazine*, 6(3):10–21, 2014.

[9] S. Amarakoon, J. Smith, and B. Segal. Application of life-cycle assessment to nanoscale technology: Lithium-ion batteries for electric vehicles. Technical Report EPA 744-R-12-00, U.S. Environmental Protection Agency, Washington, D.C., USA, 2013.

[10] F.A. Amoroso and G. Cappuccino. Potentiality of variable-rate PEVs charging strategies for smart grids. In *Proc. IEEE PowerTech*, pages 1–6, Trondheim, Norway, June 2011.

[11] D. Angeli and P.-A. Kountouriotis. A stochastic approach to dynamic-demand refrigerator control. *IEEE Transactions on Control Systems Technology*, 20(3):581–592, May 2012.

[12] O. Ardakanian, C. Rosenberg, and S. Keshav. Realtime distributed congestion control for electrical vehicle charging. *ACM SIGMETRICS Performance Evaluation Review*, 40(3):38–42, December 2012.

[13] J. Axsen, A. Burke, and K. Kurani. Batteries for plug-in hybrid electric vehicles (PHEVs): Goals and the state of technology circa 2008. Report UCD-ITS-RR-08-14, Institute of Transportation Studies, University of California, Davis, CA, USA, 2008.

[14] P. Ball. Lithium air batteries: Up in the air. Press release, IBM, 2012. Available online at https://www.zurich.ibm.com/pdf/news/lithium_air_batteries.pdf, last accessed: July 2017.

[15] S. Batmunkh, T.S. Tseyen-Oidov, and Z. Battogtokh. Survey to develop standards on air polluting emissions from the power plants. In *Proc. IEEE Third International Forum on Strategic Technologies (IFOST)*, pages 615–619, Novosibirsk, Russia, June 2008.

[16] R. Bellman. On a routing problem. *Quarterly of Applied Mathematics*, 1(16):87–90, 1958.

[17] A. Berman, C. King, and R. Shorten. A characterization of common diagonal stability over cones. *Linear and Multilinear Algebra*, 60(10):1117–1123, 2012.

[18] A. Berman and R. Plemmons. *Nonnegative Matrices in the Mathematical Sciences*. SIAM, Philadelphia, PA, USA, 1994 (Revised edition of 1979 original).

[19] D.P. Bertsekas. A simple and fast label correcting algorithm for shortest paths. *Networks*, 23:703–709, 1993.

[20] P.G. Boulter, T.J. Barlow, and I.S. McCrae. Emission factors 2009: Report 3-exhaust emission factors for road vehicles in the United Kingdom. *TRL Report PPR356. TRL Limited, Wokingham, UK*, 2009. Available online at https://www.gov.uk/government/uploads/system/

uploads/attachment_data/file/4249/report-3.pdf, last accessed: July 2017.

[21] S. Boyd, N. Parikh, E. Chu, B. Peleato, and J. Eckstein. Distributed optimization and statistical learning via the alternating direction method of multipliers. *Foundations and Trends® in Machine Learning*, 3(1):1–122, 2011.

[22] S. Boyd and L. Vanderberghe. *Convex Optimization*. Cambridge University Press, New York, NY, USA, 2004.

[23] J. Brady and M. O'Mahony. Travel to work in Dublin. The potential impacts of electric vehicles on climate change and urban air quality. *Transportation Research Part D*, 16(2):188–193, 2011.

[24] M. Brenna, F. Foiadelli, M. Longo, and D. Zaninelli. e-Mobility forecast for the transnational e-Corridor planning. *IEEE Transactions on Intelligent Transportation Systems*, 17(3):680–689, 2016.

[25] A. Brooks, E. Lu, D. Reicher, C. Spirakis, and B. Weihl. Demand dispatch. *IEEE Power and Energy Magazine*, 8(3):20–29, May - June 2010.

[26] A. Buchenscheit, F. Schaub, F. Kargl, and M. Weber. A VANET-based emergency vehicle warning system. In *Proc. IEEE Vehicular Networking Conference (VNC)*, pages 1–8, Tokyo, Japan, October 2009.

[27] K. Bullis. How improved batteries will make electric vehicles competitive. *MIT Technology Review*, November 2012. Available online at https://www.technologyreview.com/s/506881/how-improved-batteries-will-make-electric-vehicles-competitive/, last accessed: July 2017.

[28] K. Bullis. Will fast charging make electric vehicles practical? *MIT Technology Review*, September 2012. Available online at https://www.technologyreview.com/s/429283/will-fast-charging-make-electric-vehicles-practical/, last accessed: July 2017.

[29] D.S. Callaway and I.A. Hiskens. Achieving controllability of electric loads. *Proceedings of the IEEE*, 99(1):184–199, January 2011.

[30] Y. Cao, N. Wang, Y.J. Kim, and C. Ge. A reservation based charging management for on-the-move EV under mobility uncertainty. In *Proc. IEEE Online Conference on Green Communications (Online-GreenComm)*, pages 11–16, Piscataway, NJ, USA, November 2015.

[31] L. Carradore and R. Turri. Electric vehicles participation in distribution network voltage regulation. In *Proc. 45th IEEE International Universities Power Engineering Conference (UPEC)*, pages 1–6, Cardiff, United Kingdom, August 2010.

[32] O. Carsten, F. Lai, K. Chorlton, P. Goodman, D. Carslaw, and S. Hess. Speed limit adherence and its effect on road safety and climate change, final report. In *Technical Report*, University of Leeds, Institute for Transport Studies, October 2008.

[33] H. Chen, J.C. Kwong, R. Copes, K. Tu, P.J. Villeneuve, A. van Donkelaar, P. Hystad, R.V. Martin, B.J. Murray, B. Jessiman, A.S. Wilton, A. Kopp, and R.T. Burnett. Living near major roads and the incidence of dementia, Parkinson's disease, and multiple sclerosis: a population-based cohort study. *The Lancet*, 389(10070):718–726, 2017.

[34] D.-M. Chiu and R. Jain. Analysis of the increase and decrease algorithms for congestion avoidance in computer networks. *Computer Networks and ISDN Systems*, 17(1):1–14, 1989.

[35] D. Christen, S. Tschannen, and J. Biela. Highly efficient and compact DC-DC converter for ultra-fast charging of electric vehicles. In *Proc. 15th IEEE International Power Electronics and Motion Control Conference*, pages LS5d.3–1–LS5d.3–8, Novi Sad, Serbia, September 2012.

[36] K. Clement, E. Haesen, and J. Driesen. Coordinated charging of multiple plug-in hybrid electric vehicles in residential distribution grids. In *Proc. IEEE PES Power Systems Conference and Exposition (PSCE)*, pages 1–7, Seattle, WA, USA, March 2009.

[37] K. Clement-Nyns, E. Haesen, and J. Driesen. The impact of charging plug-in hybrid electric vehicles on a residential distribution grid. *IEEE Transactions on Power Systems*, 25(1):371–380, Feb 2010.

[38] T. Cone. California air pollution kills more people than car crashes, study shows. *Article in The Huffington Post*, 2008.

[39] M. Corless, C. King, R. Shorten, and F. Wirth. *AIMD Dynamics and Distributed Resource Allocation*, volume 29 of *Advances in Design and Control*. SIAM, Philadelphia, PA, USA, 2016.

[40] E. Crisostomi, S. Kirkland, and R. Shorten. A Google-like model of road network dynamics and its application to regulation and control. *International Journal of Control*, 84(3):633–651, 2011.

[41] E. Crisostomi, S. Kirkland, and R. Shorten. Markov Chain based emissions models: a precursor for green control. In J.H. Kim and M.J. Lee, editors, *Green IT: Technologies and Applications*, pages 381–400. Springer, Heidelberg, 2011.

[42] E. Crisostomi, S. Kirkland, and R. Shorten. Robust and risk-averse routing algorithms in road networks. In *Proc. IFAC World Congress*, pages 9512–9517, Milan, Italy, August 2011.

[43] E. Crisostomi, M. Liu, M. Raugi, and R. Shorten. Plug-and-play distributed algorithms for optimised power generation in a microgrid. *IEEE Transactions on Smart Grid*, 5(4):2145–2154, 2014.

[44] E. Crisostomi, R. Shorten, and F. Wirth. Smart cities: A golden age for control theory? [industry perspective]. *IEEE Technology and Society Magazine*, 35(3):23–24, 2016.

[45] G.B. Dantzig. Discrete-variable extremum problems. *Operations Research*, 5(2):266–277, 1957.

[46] J. Davies and K.S. Kurani. Recharging behavior of households' plug-in hybrid electric vehicles. *Transportation Research Record*, 2191:75–83, 2010.

[47] J. Davies and K.S. Kurani. Moving from assumption to observation: Implications for energy and emissions impacts of plug-in hybrid electric vehicles. *Energy Policy*, 62:550–560, 2013.

[48] J. de Hoog, T. Alpcan, M. Brazil, D.A. Thomas, and I. Mareels. Optimal charging of electric vehicles taking distribution network constraints into account. *IEEE Transactions on Power Systems*, 30(1):365–375, 2015.

[49] R.P.R. de Oliveira, S. Vaughan, and J.G. Hayes. Range estimation for the Nissan Leaf and Tesla Roadster using simplified power train models. In *Proc. 2011 Irish Transport Research Network Conference*, pages 1–7. University College Cork, Ireland, August 2011.

[50] M.M. de Weerdt, S. Stein, and E.H. Gerding. Intention-aware routing of electric vehicles. *IEEE Transactions on Intelligent Transportation Systems*, 17(5):1472–1482, 2016.

[51] S. Deilami, A.S. Masoum, P.S. Moses, and M.A.S. Masoum. Real-time coordination of plug-in electric vehicle charging in smart grids to minimize power losses and improve voltage profile. *IEEE Transactions on Smart Grid*, 2(3):456–467, 2011.

[52] E.W. Dijkstra. A note on two problems in connexion with graphs. *Numerische Mathematik*, 1:269–271, 1959.

[53] F. Dion, J.-S. Oh, and R. Robinson. Virtual testbed for assessing probe vehicle data in IntelliDrive systems. *IEEE Transactions on Intelligent Transportation Systems*, 12(3):635–644, 2011.

[54] S. Dirks, C. Gurdgiev, and M. Keeling. Smarter cities for smarter growth. *IBM Institute for Business Value, White Paper*, pages 1–24, 2010.

[55] The Economist. The last Kodak moment? *The Economist Print Edition, Business section*, January 2012. Available online at `http://www.economist.com/node/21542796`, last accessed: July 2017.

[56] The Economist. Seeing the back of the car. *The Economist Print Edition, Briefing section*, September 2012. Available online at `www.economist.com/node/21563280?frsc=dg\%7Ca`, last accessed: July 2017.

[57] M. Faizrahnemoon, A. Schlote, L. Maggi, E. Crisostomi, and R. Shorten. A big-data model for multi-modal public transportation with application to macroscopic control and optimisation. *International Journal of Control*, 88(11):2354–2368, 2015.

[58] P. Fan, B. Sainbayar, and S. Ren. Operation analysis of fast charging stations with energy demand control of electric vehicles. *IEEE Transactions on Smart Grid*, 6(4):1819–1826, 2015.

[59] Z. Fan. Distributed charging of PHEVs in a smart grid. In *Proc. IEEE International Conference on Smart Grid Communications (SmartGridComm)*, pages 255–260, Brussels, Belgium, October 2011.

[60] Z. Fan. Distributed demand response and user adaptation in smart grids. In *Proc. IFIP/IEEE International Symposium on Integrated Network Management*, pages 726–729, Dublin, Ireland, May 2011.

[61] R. Farrington and J. Rugh. Impact of vehicle air-conditioning on fuel economy, tailpipe emissions, and electric vehicle range: Preprint. In *Earth Technologies Forum*, pages 1–10, Washington, D.C., USA, October 2000.

[62] H. Farzin, M. Fotuhi-Firuzabad, and M. Moeini-Aghtaie. A practical scheme to involve degradation cost of lithium-ion batteries in Vehicle-to-Grid applications. *IEEE Transactions on Sustainable Energy*, 7(4):1730–1738, 2016.

[63] L.P. Fernández, T.G. San Román, R. Cossent, C.M. Domingo, and P. Frías. Assessment of the impact of plug-in electric vehicles on distribution networks. *IEEE Transactions on Power Systems*, 26(1):206–213, 2011.

[64] P. Finn, C. Fitzpatrick, and M. Leahy. Increased penetration of wind generated electricity using real time pricing and demand side management. In *Proc. IEEE International Symposium on Sustainable Systems and Technology (ISSST)*, pages 1–6, Tempe, AZ, USA, May 2009.

[65] A.R. Fioravanti, J. Mareček, R.N. Shorten, M. Souza, and F.R. Wirth. On classical control and smart cities. *Submitted for publication*, 2017. Available online at `https://arxiv.org/pdf/1703.07308.pdf`, last accessed: July 2017.

[66] Organisation for Economic Co-operation and Development (OECD). The cost of air pollution: Health impacts of road transport. *OECD Publishing*, 2014.

[67] G. Froyland. Extracting dynamical behavior via Markov models. In Alistair I. Mees, editor, *Nonlinear Dynamics and Statistics*, pages 281–321. Birkhäuser, Boston, 2001.

[68] L. Gan, U. Topcu, and S.H. Low. Optimal decentralized protocol for electric vehicle charging. *IEEE Transactions on Power Systems*, 28(2):940–951, 2013.

[69] Y. Geng and C.G. Cassandras. New "Smart Parking" system based on resource allocation and reservations. *IEEE Transactions on Intelligent Transportation Systems*, 14(3):1129–1139, 2013.

[70] V. Ghanbari, P. Wu, and P.J. Antsaklis. Large-scale dissipative and passive control systems and the role of star and cyclic symmetries. *IEEE Transactions on Automatic Control*, 61(11):3676–3680, 2016.

[71] M. Gharbaoui, B. Martini, R. Bruno, L. Valcarenghi, M. Conti, and P. Castoldi. Designing and evaluating activity-based electric vehicle charging in urban areas. In *Proc. IEEE International Electric Vehicle Conference (IEVC)*, pages 1–5, Silicon Valley, CA, USA, October 2013.

[72] E. Giordano, R. Frank, G. Pau, and M. Gerla. CORNER: a realistic urban propagation model for VANET. In *Proc. 7th IEEE International Conference on Wireless On-demand Network Systems and Services (WONS)*, pages 57–60, Kranjska Gora, Slovenia, February 2010.

[73] Q. Gong, S. Midlam-Mohler, V. Marano, and G. Rizzoni. Study of PEV charging on residential distribution transformer life. *IEEE Transactions on Smart Grid*, 3(1):404–412, March 2012.

[74] S. Grammatico, F. Parise, M. Colombino, and J. Lygeros. Decentralized convergence to Nash equilibria in constrained deterministic mean field control. *IEEE Transactions on Automatic Control*, 61(11):3315–3329, 2016.

[75] M. Granovskii, I. Dincer, and M.A. Rosen. Economic and environmental comparison of conventional, hybrid, electric and hydrogen fuel cell vehicles. *Journal of Power Sources*, 159(2):1186–1193, 2006.

[76] W. Griggs, J.Y. Yu, F. Wirth, F. Häusler, and R. Shorten. On the design of campus parking systems with QoS guarantees. *IEEE Transactions on Intelligent Transportation Systems*, 17(5):1428–1437, 2016.

[77] W.M. Griggs, R.H. Ordóñez Hurtado, E. Crisostomi, F. Häusler, K. Massow, and R.N. Shorten. A large-scale SUMO-based emulation

platform. *IEEE Transactions on Intelligent Transportation Systems*, 16(6):3050–3059, 2015.

[78] C.M. Grinstead and J.L. Snell. *Introduction to Probability*. American Mathematical Society (2nd Edition), Providence, RI, USA, 2003.

[79] N. Groot, B. De Schutter, and H. Hellendoorn. Toward system-optimal routing in traffic networks: A reverse Stackelberg game approach. *IEEE Transactions on Intelligent Transportation Systems*, 16(1):29–40, 2015.

[80] Y. Gu, F. Häusler, W. Griggs, E. Crisostomi, and R. Shorten. Smart procurement of naturally generated energy SPONGE for PHEVs. *International Journal of Control*, 89(7):1467–1480, 2016.

[81] Y. Gu, M. Liu, J. Naoum-Sawaya, E. Crisostomi, G. Russo, and R. Shorten. Pedestrian-aware engine management strategies for plug-in hybrid electric vehicles. *IEEE Transactions on Intelligent Transportation Systems, accepted for publication, available online at https: // arxiv. org/ abs/ 1612. 07568*, 2017.

[82] A. Gusrialdi, Z. Qu, and M.A. Simaan. Scheduling and cooperative control of electric vehicles charging at highway service stations. In *Proc. 53rd IEEE Conference on Decision and Control (CDC)*, pages 6465–6471, Los Angeles, CA, USA, December 2014.

[83] K. Gyimesi and R. Viswanathan. The shift to electric vehicles: Putting consumers in the driver's seat. Executive report, IBM Global Business Services, Somers, New York, USA, 2011.

[84] M. Haklay and P. Weber. Openstreetmap: User-generated street maps. *IEEE Pervasive Computing*, 7(4):12–18, 2008.

[85] A. Harris. Charge of the electric car. *IET Engineering and Technology*, 4(10):52–53, June 2009.

[86] F. Häusler. *Topics in Feedback Control for ITS Applications in Smart Cities*. PhD Thesis, Technische Universität Berlin, 2016.

[87] F. Häusler, E. Crisostomi, A. Schlote, I. Radusch, and R. Shorten. Stochastic park-and-charge balancing for fully electric and plug-in hybrid vehicles. *IEEE Transactions on Intelligent Transportation Systems*, 15(2):895–901, 2014.

[88] F. Häusler, M. Faizrahnemoon, E. Crisostomi, A. Schlote, I. Radusch, and R. Shorten. A framework for real-time emissions trading in large scale vehicle fleets. *IET Intelligent Transport Systems*, 9(3):275–284, 2015.

[89] Y. He, B. Venkatesh, and L. Guan. Cost minimization for charging and discharging of electric vehicles. In *Proc. IEEE International Conference on Smart Grid Communications (SmartGridComm)*, pages 273–278, Brussels, Belgium, October 2011.

[90] D. Helbing and B.A. Huberman. Coherent moving states in highway traffic. *Nature*, 396:738–740, 1998.

[91] A. Herrmann, M. Liu, and R. Shorten. A new take on protecting cyclists in smart cities. *Submitted for publication*, pages 1–6, 2017. Available online at `https://arxiv.org/abs/1704.04540`, last accessed: July 2017.

[92] B.-M. Hodge, D. Lew, M. Milligan, H. Holttinen, S. Sillanpää, E. Gómez Lázaro, R. Scharff, L. Söder, X.G. Larsén, G. Giebel, D. Flynn, and J. Dobschinski. Wind power forecasting error distributions: An international comparison. In *Proc. 11th Annual International Workshop on Large-Scale Integration of Wind Power into Power Systems as well as on Transmission Networks for Offshore Wind Power Plants*, pages 1–7, Lisbon, Portugal, November 2012.

[93] W. Hoeffding. Probability inequalities for sums of bounded random variables. *Journal of the American Statistical Association*, 58(301):13–30, 1963.

[94] R. Horn and C. Johnson. *Matrix Analysis*. Cambridge University Press, Cambridge, UK, 1985.

[95] N.B. Hounsell, B.P. Shrestha, J. Piao, and M. McDonald. Review of urban traffic management and the impacts of new vehicle technologies. *IET Intelligent Transport Systems*, 3(4):419–428, 2009.

[96] E.B. Iversen, J.K. Møller, J.M. Morales, and H. Madsen. Inhomogeneous Markov models for describing driving patterns. *IEEE Transactions on Smart Grid*, 8(2):581–588, 2017.

[97] R.D. Jong, M. Ahman, R. Jacobs, and E. Dumitrescu. Hybrid electric vehicles: An overview of current technology and its application in developing and transitional countries. *United Nations Environment Progamme (UNEP), Nairobi, Kenya*, 2009.

[98] T. Jurik, A. Cela, R. Hamouche, R. Natowicz, A. Reama, S.-I. Niculescu, and J. Julien. Energy optimal real-time navigation system. *IEEE Intelligent Transportation Systems Magazine*, 6(3):66–79, 2014.

[99] M. Kahlen, K. Valogianni, W. Ketter, and J.V. Dalen. A profitable business model for electric vehicle fleet owners. In *Proc. IEEE International Conference on Smart Grid Technologies, Economics and Policies*, pages 1–5, Nuremberg, Germany, December 2012.

[100] E.L. Karfopoulos, K.A. Panourgias, and N.D. Hatziargyriou. Distributed coordination of electric vehicles providing V2G regulation services. *IEEE Transactions on Power Systems*, 31(4):2834–2846, 2016.

[101] F.P. Kelly. Mathematical modelling of the Internet. In *Proc. 4th International Congress on Industrial and Applied Mathematics (ICIAM)*, pages 105–116, Edinburgh, UK, July 1999.

[102] J.G. Kemeny and J.L. Snell. *Finite Markov Chains*. Van Nostrand, Princeton, NJ, USA, 1960.

[103] S. Keshav and C. Rosenberg. Direct adaptive control of electricity demand, September 2010. Technical Report CS-2010-17, Cheriton School of Computer Science, University of Waterloo, Canada.

[104] H.K. Khalil. *Nonlinear Systems*. Prentice Hall (3rd Edition), London, UK, 2002.

[105] M. Khodabakhshian, L. Feng, and J. Wikander. Predictive control of the engine cooling system for fuel efficiency improvement. In *Proc. IEEE International Conference on Automation Science and Engineering (CASE)*, pages 61–66, Taipei, Taiwan, August 2014.

[106] C. King, W. Griggs, F. Wirth, K. Quinn, and R. Shorten. Alleviating a form of electric vehicle range anxiety through on-demand vehicle access. *International Journal of Control*, 88(4):717–728, 2015.

[107] C. King, W. Griggs, F. Wirth, and R. Shorten. Using a car sharing model to alleviate electric vehicle range anxiety. In *Proc. Sixteenth Yale Workshop on Adaptive and Learning Systems*, New Haven, CT, USA, June 2013. Yale University.

[108] B. Kirby and E. Hirst. Load as a resource in providing ancillary services. In *Proc. American Power Conference*, volume 61, Chicago, IL, USA, April 1999.

[109] F. Knorn, M. Corless, and R. Shorten. A result on implicit consensus with application to emissions control. In *Proc. IEEE Control and Decision Conference (CDC)*, pages 1299–1304, Orlando, Fl, USA, December 2011.

[110] F. Knorn, M. Corless, and R. Shorten. Results in cooperative control and implicit consensus. *International Journal of Control*, 84(3):476–495, 2011.

[111] F. Knorn, R. Stanojevic, M. Corless, and R. Shorten. A framework for decentralised feedback connectivity control with application to sensor networks. *International Journal of Control*, 82(11):2095–2114, 2009.

[112] K. Kostková, L. Omelina, P. Kyčina, and P. Jamrich. An introduction to load management. *Electric Power Systems Research*, 95:184–191, 2013.

[113] D. Krajzewicz, J. Erdmann, M. Behrisch, and L. Bieker. Recent development and applications of SUMO – Simulation of Urban MObility. *International Journal on Advances in Systems and Measurements*, 5(3):128–138, 2012.

[114] A.N. Langville and C.D. Meyer. *Google's PageRank and Beyond - The Science of Search Engine Rankings*. Princeton University Press, Princeton, NJ, USA, 2006.

[115] V. Larsson, J. Johannesson Mårdh, B. Egardt, and S. Karlsson. Commuter route optimized energy management of hybrid electric vehicles. *IEEE Transactions on Intelligent Transportation Systems*, 15(3):1145–1154, 2014.

[116] L. Li, B. Yan, C. Yang, Y. Zhang, Z. Chen, and G. Jiang. Application oriented stochastic energy management for plug-in hybrid electric bus with AMT. *IEEE Transactions on Vehicular Technology*, 65(6):4459–4470, 2016.

[117] L. Li, C. Yang, Y. Zhang, L. Zhang, and J. Song. Correctional DP-based energy management strategy of plug-in hybrid electric bus for city-bus route. *IEEE Transactions on Vehicular Technology*, 64(7):2792–2803, 2015.

[118] N. Li, L. Chen, and S.H. Low. Optimal demand response based on utility maximization in power networks. In *Proc. IEEE PES General Meeting*, pages 1–8, Detroit, MI, USA, July 2011.

[119] D. Linden and T.B. Reddy. *Handbook of Batteries*. McGraw-Hill, 2002. 3rd ed.

[120] H. Liu, L. Zhang, D. Sun, and D. Wang. Optimize the settings of variable speed limit system to improve the performance of freeway traffic. *IEEE Transactions on Intelligent Transportation Systems*, 16(6):3249–3257, 2015.

[121] M. Liu, W. Griggs, C. King, F. Wirth, P. Borrel, and R. Shorten. Applying a QoS-based fleet dimension method to reduce fleet emissions. In *Proc. IEEE International Conference on Connected Vehicles and Expo (ICCVE)*, pages 732–733, Las Vegas, NV, USA, December 2013.

[122] M. Liu, R.H. Ordóñez Hurtado, F. Wirth, Y. Gu, E. Crisostomi, and R. Shorten. A distributed and privacy-aware speed advisory system for optimizing conventional and electric vehicle networks. *IEEE Transactions on Intelligent Transportation Systems*, 17(5):1308–1318, 2016.

[123] M. Liu, R.H. Ordóñez Hurtado, F.R. Wirth, Y. Gu, E. Crisostomi, and R. Shorten. An intelligent speed advisory system for electric vehicles. In *Proc. IEEE International Conference on Connected Vehicles and Expo (ICCVE)*, pages 84–88, Shenzhen, China, October 2015.

[124] M. Liu, F. Wirth, M. Corless, and R. Shorten. On the stability and convergence of a class of consensus systems with a nonlinear input. *Automatica, provisionally accepted for publication, available online at http://arxiv.org/abs/1506.01430, last accessed: July 2017*, 2015.

[125] S. Lobach. *Adaptive urban emission control by online rerouting*. PhD Thesis, Technische Universität Berlin, 2013.

[126] M. Lowe, S. Tokuoka, T. Trigg, and G. Gereffi. Lithium-ion batteries for electric vehicles: The U.S. value chain. Research report, Center on Globalization, Governance & Competitiveness, Duke University, Durham, NC, USA, October 2010.

[127] D. Lowell. Comparison of modern CNG, Diesel and Diesel hybrid-electric transit buses: Efficiency & environmental performance. *Report of M.J. Bradley & Associates, LLC*, 2013.

[128] J. Macedo, Z. Kokkinogenis, G. Soares, D. Perrotta, and R.J.F. Rossetti. A HLA-based multi-resolution approach to simulating electric vehicles in simulink and SUMO. In *Proc. 16th IEEE International Annual Conference on Intelligent Transportation Systems*, pages 2367–2372, The Hague, Netherlands, October 2013.

[129] D. MacKay. *Sustainable Energy - Without the Hot Air*. UIT Cambridge Ltd, 2008.

[130] R. Maia, M. Silva, R. Araújo, and U. Nunes. Electric vehicle simulator for energy consumption studies in electric mobility systems. In *Proc. IEEE Forum on Integrated and Sustainable Transportation Systems*, pages 227–232, Vienna, Austria, June 2011.

[131] R. Mangharam, D. Weller, R. Rajkumar, P. Mudalige, and F. Bai. Groovenet: a hybrid simulator for vehicle-to-vehicle networks. In *Proc. 3rd IEEE Annual International Conference on Mobile and Ubiquitous Systems*, pages 1–8, San Jose, CA, USA, July 2006.

[132] K. Mets, T. Verschueren, F. De Turck, and C. Develder. Evaluation of multiple design options for smart charging algorithms. In *Proc. IEEE International Conference on Communications (ICC)*, pages 1–5, Kyoto, Japan, June 2011.

[133] V. Moosavi and L. Hovestadt. Modeling urban traffic dynamics in co-existence with urban data streams. In *Proc. 2nd ACM SIGKDD International Workshop on Urban Computing*, pages 1–7, Chicago, IL, USA, August 2013.

[134] L. Moreau. Stability of multiagent systems with time-dependent communication links. *IEEE Transactions on Automatic Control*, 50(2):169–182, 2005.

[135] T. Morimura, T. Osogami, and T. Idé. *Solving inverse probem of Markov chain with partial observations*. IBM Research Report RT0952, 2013.

[136] P.J. Moylan and D.J. Hill. Stability criteria for large-scale systems. *IEEE Transactions on Automatic Control*, 23(2):143–149, 1978.

[137] J. Naoum-Sawaya, E. Crisostomi, M. Liu, Y. Gu, and R. Shorten. Smart procurement of naturally generated energy (SPONGE) for plug-in hybrid electric buses. *IEEE Transactions on Automation Science and Engineering*, 14(2):598–607, 2017.

[138] M. Neaimeh, G.A. Hill, Y. Hübner, and P.T. Blythe. Routing systems to extend the driving range of electric vehicles. *IET Intelligent Transport Systems*, 7(3):327–336, 2013.

[139] T. Niknam and A. Kavousifard. Impact of thermal recovery and hydrogen production of fuel cell power plants on distribution feeder reconfiguration. *IET Generation, Transmission & Distribution*, 6(9):831–843, September 2012.

[140] M. Nilsson. Electric vehicles: The phenomenon of range anxiety. Public Deliverable FP7-ICT-2009-4-249105, Lindholmen Science Park, Sweden, 2011.

[141] J.R. Norris. *Markov Chains*. Cambridge Series in Statistical and Probabilistic Mathematics. Cambridge University Press, Cambridge, UK, 1998.

[142] K. Pandit, D. Ghosal, H.M. Zhang, and C.-N. Chuah. Adaptive traffic signal control with vehicular ad hoc networks. *IEEE Transactions on Vehicular Technology*, 62(4):1459–1471, 2013.

[143] P. Papadopoulos, S. Skarvelis-Kazakos, I. Grau, L.M. Cipcigan, and N. Jenkins. Predicting electric vehicle impacts on residential distribution networks with distributed generation. In *Proc. IEEE Vehicle Power and Propulsion Conference (VPPC)*, pages 1–5, Lille, France, September 2010.

[144] Y.B. Park and S.J. Yoon. A decision support system for the operation of vending machine supply chains with product substitution and varying vehicle speed. *Applied Mechanics and Materials*, 284:3617–3621, 2013.

[145] M. Pavone, E. Frazzoli, and F. Bullo. Adaptive and distributed algorithms for vehicle routing in a stochastic and dynamic environment. *IEEE Transactions on Automatic Control*, 56(6):1259–1274, 2011.

[146] J.A. Peças Lopes, F.J. Soares, and P.M. Rocha Almeida. Identifying management procedures to deal with connection of electric vehicles in the grid. In *Proc. IEEE Bucharest Power Tech Conference*, pages 1–8, Bucharest, Romania, June 2009.

[147] J.A. Peças Lopes, F.J. Soares, and P.M. Rocha Almeida. Integration of electric vehicles in the electric power system. *Proceedings of the IEEE*, 99(1):168–183, 2011.

[148] J.L.F. Pereira and R.J.F. Rossetti. An integrated architecture for autonomous vehicles simulation. In *Proc. 27th Annual ACM Symposium on Applied Computing*, pages 286–292, Riva del Garda, Italy, March 2012.

[149] V. Punzo and B. Ciuffo. Integration of driving and traffic simulation: issues and first solutions. *IEEE Transactions on Intelligent Transportation Systems*, 12(2):354–363, 2011.

[150] G.A. Putrus, P. Suwanapingkarl, D. Johnston, E.C. Bentley, and M. Narayana. Impact of electric vehicles on power distribution networks. In *Proc. IEEE Vehicle Power and Propulsion Conference (VPPC)*, pages 827–831, Dearborn, MI, USA, September 2009.

[151] K. Qian, C. Zhou, M. Allan, and Y. Yuan. Modeling of load demand due to EV battery charging in distribution systems. *IEEE Transactions on Power Systems*, 26(2):802–810, May 2011.

[152] M. Quinlan, T.-C. Au, J. Zhu, N. Stiurca, and P. Stone. Bringing simulation to life: a mixed reality autonomous intersection. In *Proc. IEEE/RSJ International Conference on Intelligent Robots and Systems*, pages 6083–6088, Taipei, Taiwan, October 2010.

[153] S. Rahim. European commission aims to phase out gasoline- and diesel-powered cars in cities by 2050. *Scientific American*, 2011. Available online at http://www.scientificamerican.com/article/european-commission-proposes-push-gas-diesel-cars-out-cities/, last accessed: July 2017.

[154] P. Richardson, D. Flynn, and A. Keane. Optimal charging of electric vehicles in low-voltage distribution systems. *IEEE Transactions on Power Systems*, 27(1):268–279, February 2012.

[155] P. Richardson, J. Taylor, D. Flynn, and A. Keane. Stochastic analysis of the impact of electric vehicles on distribution networks. In *Proc. 21st International Conference on Electricity Distribution (CIRED)*, pages 1–4, Frankfurt, Germany, June 2011.

[156] F.J. Ros, J.A. Martinez, and P.M. Ruiz. A survey on modeling and simulation of vehicular networks: Communications, mobility, and tools. *Computer Communications*, 43(1):1–15, 2014.

[157] A. Y. Saber and G. K. Venayagamoorthy. Resource scheduling under uncertainty in a smart grid with renewables and plug-in vehicles. *IEEE Systems Journal*, 6(1):103–109, March 2012.

[158] A. Sakti. Factors that can contribute to cost reduction of lithium-ion batteries for personal vehicles. Fact sheet, American Council for an Energy-Efficient Economy (ACEE), Washington, D.C., USA, 2011.

[159] D. Samoilenko and H.M. Cho. Improvement of combustion efficiency and emission characteristics of IC diesel engine operating on ESC cycle applying Variable Geometry Turbocharger (VGT) with vaneless turbine volute. *International Journal of Automotive Technology*, 14(4):521–528, 2013.

[160] S. Sayeef, S. Heslop, D. Cornforth, T. Moore, S. Percy, J.K. Ward, A. Berry, and D. Rowe. Solar intermittency: Australia's clean energy challenge, June 2012. Available online at https://publications.csiro.au/rpr/download?pid=csiro:EP121914&dsid=DS1, last accessed: July 2017.

[161] A. Schlote. *New perspectives on modelling and control for next generation intelligent transport systems*. PhD thesis, Hamilton Institute, National University of Ireland Maynooth (NUIM), 2014.

[162] A. Schlote, E. Crisostomi, S. Kirkland, and R. Shorten. Traffic modelling framework for electric vehicles. *International Journal of Control*, 85(7):880–897, 2012.

[163] A. Schlote, E. Crisostomi, and R Shorten. Balanced routing for EVs. In *Proc. IEEE International Conference on Connected Vehicles and Expo (ICCVE)*, pages 343–344, Beijing, China, December 2012.

[164] A. Schlote, F. Häusler, T. Hecker, A. Bergmann, E. Crisostomi, I. Radusch, and R. Shorten. Cooperative regulation and trading of emissions using plug-in hybrid vehicles. *IEEE Transactions on Intelligent Transportation Systems*, 4(14):1572–1585, 2013.

[165] A. Schlote, C. King, E. Crisostomi, and R. Shorten. Delay-tolerant stochastic algorithms for parking space assignment. *IEEE Transactions on Intelligent Transportation Systems*, 15(5):1922–1935, 2014.

[166] R. Sioshansi and P. Denholm. Emissions impacts and benefits of plug-in hybrid electric vehicles and vehicle-to-grid services. *Environmental Science & Technology*, 43(4):1199–1204, 2009.

[167] R. Srikant. *The Mathematics of Internet Congestion Control*. Systems & Control: Foundations & Applications, Birkhäuser, 2003.

[168] R. Stanojević and R. Shorten. Generalized distributed rate limiting. In *Proc. 17th IEEE International Workshop on Quality of Service (IWQoS)*, pages 1–9, Charleston, SC, USA, July 2009.

[169] S. Stüdli. *Distributed load management supporting power injection and reactive power balancing.* PhD thesis, The University of Newcastle, 2015.

[170] S. Stüdli, E. Crisostomi, R. Middleton, J. Braslavsky, and R. Shorten. Distributed load management using additive increase multiplicative decrease based techniques. In Sumedha Rajakaruna, Farhad Shahnia, and Arindam Ghosh, editors, *Plug In Electric Vehicles in Smart Grids*, Power Systems, pages 173–202. Springer Singapore, 2015.

[171] S. Stüdli, E. Crisostomi, R. Middleton, and R. Shorten. AIMD-like algorithms for charging electric and plug-in hybrid vehicles. In *Proc. IEEE International Electric Vehicle Conference (IEVC)*, pages 1–8, Greenville, SC, USA, March 2012.

[172] S. Stüdli, E. Crisostomi, R. Middleton, and R. Shorten. A flexible distributed framework for realising electric and plug-in hybrid vehicle charging policies. *International Journal of Control*, 85(8):1130–1145, 2012.

[173] S. Stüdli, E. Crisostomi, R. Middleton, and R. Shorten. Optimal real-time distributed V2G and G2V management of electric vehicles. *International Journal of Control*, 87(6):1153–1162, 2014.

[174] S. Stüdli, W. Griggs, E. Crisostomi, and R. Shorten. On optimality criteria for V2G charging strategies. In *Proc. IEEE International Conference on Connected Vehicles and Expo (ICCVE)*, pages 345–346, Beijing, China, December 2012.

[175] S. Stüdli, W. Griggs, E. Crisostomi, and R. Shorten. On optimality criteria for reverse charging of electric vehicles. *IEEE Transactions on Intelligent Transportation Systems*, 15(1):451–456, 2014.

[176] S. Stüdli, R.H. Middleton, and J.H. Braslavsky. A fixed-structure automaton for load management of electric vehicles. In *Proc. IEEE European Control Conference (ECC)*, pages 3566–3571, Zurich, Switzerland, July 2013.

[177] W. Su and M.-Y. Chow. Performance evaluation of an EDA-based large-scale plug-in hybrid electric vehicle charging algorithm. *IEEE Transactions on Smart Grid*, 3(1):308–315, 2012.

[178] S. Sweeney, R.H. Ordóñez Hurtado, F. Pilla, G. Russo, D. Timoney, and R. Shorten. Cyberphysics, pollution mitigation, and pedelecs. *Submitted to the IEEE Transactions on Intelligent Transportation Systems*, available online at *http://arxiv.org/abs/1706.00646*, 2017.

[179] X. Tang, J. Liu, X. Wang, and J. Xiong. Electric vehicle charging station planning based on weighted Voronoi diagram. In *Proc. IEEE International Conference on Transportation and Mechanical and Electrical Engineering (TMEE)*, pages 1297–1300, Changchun, China, December 2011.

[180] E.D. Tate, M.O. Harpster, and P.J. Savagian. The electrification of the automobile: From conventional hybrid, to plug-in hybrids, to extended range electric vehicles. In *SAE World Congress Exhibition, reprinted from: Advanced Hybrid Vehicle Powertrain*, pages 156–166, Detroit, MI, USA, April 2008.

[181] E. Thompson, R.H. Ordóñez Hurtado, W. Griggs, and R. Shorten. On charge point anxiety and the sharing economy. In *Submitted to the IEEE International Conference on Intelligent Transportation Systems (ITSC)*, Yokohama, Japan, October 2017.

[182] F. Tianheng, Y. Lin, G. Qing, H. Yanqing, Y. Ting, and Y. Bin. A supervisory control strategy for plug-in hybrid electric vehicles based on energy demand prediction and route preview. *IEEE Transactions on Vehicular Technology*, 64(5):1691–1700, 2015.

[183] T. Tielert, M. Killat, H. Hartenstein, R. Luz, S. Hausberger, and T. Benz. The impact of traffic-light-to-vehicle communication on fuel consumption and emissions. In *Proc. IEEE Internet of Things Conference*, pages 1–8, Tokyo, Japan, November 2010.

[184] N. Tradisauskas, J. Juhl, H. Lahrmann, and C.S. Jensen. Map matching for intelligent speed adaptation. *IET Intelligent Transport Systems*, 3(1):57–66, 2009.

[185] R. Van Haaren. Assessment of electric cars' range requirements and usage patterns based on driving behavior recorded in the national household travel survey of 2009. In *A Study conducted as part of the Solar Journey USA project*, 2011. Available online at http://www.solarjourneyusa.com/HowFarWeDrive_v1.2.pdf, last accessed: July 2017.

[186] H.-J. Wagner, M.K. Koch, J. Burkhardt, T. Große-Böckmann, N. Feck, and P. Kruse. Co2-emissionen der stromerzeugung. *BWK - Das Energie-Fachmagazin*, 59(10):44–51, 2007.

[187] Y. Wang, J. Jiang, and T. Mu. Context-aware and energy-driven route optimization for fully electric vehicles via crowdsourcing. *IEEE Transactions on Intelligent Transportation Systems*, 14(3):1331–1345, 2013.

[188] J.K. Ward, J. Wall, S. West, and R. de Dear. Beyond comfort managing the impact of HVAC control on the outside world. In *Proc.*

Windsor Conference: Air Conditioning and the Low Carbon Cooling Challenge, Cumberland Lodge, Windsor, UK, July 2008. Network for Comfort and Energy Use in Buildings. Available online at `nceub.org.uk/dokuwiki/lib/exe/fetch.php?media=nceub:uploads:members:w2008:session1:w2008_53ward.pdf`, last accessed: July 2017.

[189] A. Wegener, M. Piórkowski, M. Raya, H. Hellbrück, S. Fischer, and J-P. Hubaux. TraCI: an interface for coupling road traffic and network simulators. In *Proc. 11th Communications and Networking Simulation Symposium*, pages 155–163, Ottawa, Canada, April 2008.

[190] J. T. Wen and M. Arcak. A unifying passivity framework for network flow control. *IEEE Transactions on Automatic Control*, 49(2):162–174, 2004.

[191] Night Wind. Grid architecture for wind power production with energy storage through load shifting in refrigerated warehouses, final activity report - night wind - summary, Project no. SES6 - CT - 2006 - 20045, 2009. Available online at `http://cordis.europa.eu/documents/documentlibrary/121790181EN6.pdf`, last accessed: July 2017.

[192] F. Wirth, S. Stüdli, J.Y. Yu, M. Corless, and R. Shorten. Asynchronous algorithms for network utility maximisation with a single bit. In *Proc. IEEE European Control Conference (ECC)*, pages 291–296, Linz, Austria, July 2015.

[193] F. Wirth, S. Stüdli, J.Y. Yu, M. Corless, and R. Shorten. Nonhomogeneous place-dependent Markov Chains, unsynchronised AIMD, and network utility maximization, 2015. Available online at `http://arxiv.org/pdf/1404.5064v2.pdf`, last accessed: July 2017.

[194] J. Woodjack, D. Garas, A. Lentz, T. Turrentine, G. Tal, and M. Nicholas. Consumer perceptions and use of driving distance of electric vehicles. *Transportation Research Record*, 2287:1–8, 2012.

[195] O. Worley, D. Klabjan, and T.M. Sweda. Simultaneous vehicle routing and charging station siting for commercial Electric Vehicles. In *Proc. IEEE International Electric Vehicle Conference (IEVC)*, pages 1–3, Greenville, SC, USA, March 2012.

[196] D. Wu, D.C. Aliprantis, and L. Ying. Load scheduling and dispatch for aggregators of plug-in electric vehicles. *IEEE Transactions on Smart Grid*, 3(1):368–376, March 2012.

[197] H. Xing, M. Fu, Z. Lin, and Y. Mou. Decentralized optimal scheduling for charging and discharging of plug-in electric vehicles in smart grids. *IEEE Transactions on Power Systems*, 31(5):4118–4127, 2016.

[198] G. Yan, S. Olariu, and M.C. Weigle. Providing VANET security through active position detection. *Elsevier Computer Communications*, 31(12):2883–2897, 2008.

[199] B. Yang, M. Kaul, and C.S. Jensen. Using incomplete information for complete weight annotation of road networks. *IEEE Transactions on Knowledge and Data Engineering*, 26(5):1267–1279, 2014.

[200] H. Yang, S. Yang, Y. Xu, E. Cao, M. Lai, and Z. Dong. Electric vehicle route optimization considering time-of-use electricity price by learnable Partheno-Genetic algorithm. *IEEE Transactions on Smart Grid*, 6(2):657–666, 2015.

[201] M. Yilmaz and P.T. Krein. Review of battery charger topologies, charging power levels, and infrastructure for plug-in electric and hybrid vehicles. *IEEE Transactions on Power Electronics*, 28(5):2151–2169, 2013.

[202] W. Yuen, J. Andon, and R. Bajulaz. The Bajulaz cycle: A two-chamber internal combustion engine with increased thermal efficiency. In *SAE Technical Paper 860534*, pages 70–77, 1986.

[203] J. Zhang, B.-M. Hodge, A. Florita, S. Lu, H.F. Hamann, and V. Banunarayanan. Metrics for evaluating the accuracy of solar power forecasting. In *Proc. 3rd International Workshop on Integration of Solar Power into Power Systems*, pages 1–8, London, UK, October 2013.

[204] Y. Zhao, A. Wagh, K. Hulme, C. Qiao, A.W. Sadek, H. Xu, and L. Huang. Integrated traffic-driving-networking simulator: a unique R&D tool for connected vehicles. In *Proc. 1st IEEE International Conference on Connected Vehicles and Expo (ICCVE)*, pages 203–204, Beijing, China, December 2012.

[205] K. Zhou, J. Doyle, and K. Glover. *Robust and Optimal Control*. Prentice Hall, Upper Saddle River, NJ, USA, 1996.

Index

Additive increase, 205
Additive increase multiplicative
    decrease, *see* AIMD
ADMM, 183, 209
Advertising, 8
    charge capacity, 46
Aerodynamics
    EV power consumption, 180
Aggregation, 9, 16, 20, 94
    by-products of the ICE, 8
    emission control, *see* Control,
        aggregate emissions
    emissions, *see* Emissions,
        aggregate
    on-board equipment, 24
    pollution, *see* Pollution,
        aggregate levels
    super-batteries, 160
    vehicle emissions, 145
AIMD, 8, 131, 205
    and agent actuation, 174
    and communication
        requirements, 173
    and privacy, *see* Privacy, AIMD
    EV charging, 69, 81
    optimization, 173
    parameterization, 174
    SPONGE, 174
Air quality, 143, 144
Algorithm
    AIMD, *see* AIMD
    binary automaton, *see* EV
        charging, binary automaton
    clustering, 46
    consensus, *see* Consensus
    Dijkstra's, 35, 39
    EV charging, *see* EV charging

    stochastic, 52
Ancillary service, 58, 63, 171
Ancillary systems
    EV power consumption, 181
Angst, *see* Anxiety
Anxiety, 94, 132
    charge point, *see* Charge point,
        anxiety
    range, *see* Range anxiety
Autonomous driving, 9, 197
Auxiliary service, 8
Auxiliary systems, 35

Balancing, 43
    behavior, 49
    charging load, *see* EV charging,
        balancing
    energies, 48
    EV charging, *see* EV charging,
        balancing
    load, 15
    power grid demand, 81
    stochastic, 46
    traffic, 26
    vehicles, 40, 47
Battery, 5, 22
    as filter, *see* Filter
    capacity, 59
    cost, *see* Cost, battery
    inadequate level, 98
    lifetime, 141
    power limit, 59
    size, 5
    stresses, 63
Big-data, 20

Campus sharing, *see* Sharing, on
　　campus
Capacity constraint, 131
Capacity event, 66, 82, 206
Central management unit, *see*
　　Management, central unit
Charge point
　　anxiety, 119
Charge point
　　anxiety, 4, 94
　　availability, 4, 47
　　dynamic availability, 52
　　network, *see* Networks, charge
　　　　point
　　placing, 45
　　private, 119
　　sharing, *see* Sharing, charge
　　　　point
　　smart adapter, 134
Charging, *see* EV charging
　　fast, 98
Charging station, *see* Charge point
Collaborative routing, *see* Routing,
　　collaborative
Communication
　　AIMD, *see* AIMD, and
　　　　communication
　　　　requirements
　　overhead, 151
Congestion, 119
　　local, 40
Congestion event, *see* Capacity event
Connectivity, *see* V2V, connectivity
Consensus, 180, 182, 209
　　optimality, 185
　　optimization, 183
Consumption
　　resource, 205
Continuous-time, 205
Control, 210
　　actuation, 62
　　aggregate emissions, 151
　　architecture, 60
　　centralized, 60
　　closed-loop, 43

cooperative, 147
decentralized, 60
distributed, 48, 60
EV charging, *see also* Demand
　　side management, 59
integral, 152
open-loop, 43
theory, 107
traffic load, 26, 28
Convergence, 209
Convexity
　　of a function, 207
Cooperation, 170
Cost
　　average, 22
　　battery, 3, 86
　　charging, 57
　　depreciation, 111
　　fleet, 112
　　marginal, 99
　　overhead, 112
Cost function, 90, 130, 185, 207
Current limit, 82
Cyber-physical, 211

Data analysis, 100
Data protection, 62, 66
Delay, 43
Demand
　　power, *see* Power demand
　　synchronized, 107, 120
Demand side management, 58
DER, 57
Dijkstra's algorithm, *see* Algorithm,
　　Dijkstra's
Disruption, 7
Distributed energy resource, *see*
　　DER
Distribution
　　binomial, 105
　　exponential, 49
　　grid, *see* Power grid
　　normal, 127
　　stationary, 18, 50
Distribution grid, *see also* Power grid

Drag coefficient, 180
Drivetrain
    EV power consumption, 180
Driving behavior, *see also* Mobility,
    patterns, 159, 197
Driving patterns, *see* Mobility,
    patterns

Ecodriving, 159, 160
Electric charge point, *see* Charge
    point
Electric vehicle, *see* EV
Electrical load pattern, 57
Emissions, 2, 13, 88, 99, 144, 148,
    170, 179, 201
    aggregate, 144
    electromagnetic, 5
    fleet, 97, 116
Energy
    ancillary services, 171
    budget, 162, 169
    consumption, 13, 14, 162, 168,
        171, 179, 181
    management, *see* Management,
        energy
    regulation, 162
    storage, 93
Environment, 86
    impact, 86, 160
Environmental impact, *see*
    Environment, impact
Equality constraint, 162
Ergodicity, 184
EV, 1, 81
    charging, *see* EV charging
    full, 2, 81
    hybrid, 1, 7
    network, *see also* Networks, EV
    plug-in, 1, 2
    plug-in bus, 168
    plug-in hybrid, 1
    power consumption, *see* Power
        consumption, EV
    range, 4, 87
    routing, 35

routing , *see also* Routing, EV
EV charging, *see also* Demand side
    management, 2, 4, 15, 57,
        87, 161
    control architectures, 60
    AIMD, *see* AIMD, EV charging
    ancillary services, *see* Ancillary
        service
    balancing, 45, *see also*
        Balancing, 46
    binary automaton, 67
    communication requirements, 62
    control actuation, 62
    control methods, 63
    domestic, 70
    driver behavior, 104
    fairness, *see* Fairness, EV
        charging
    fast, 3, 14
    home unit, 133
    on-off, 67
    overnight, 4, 15, 98, 100, 116
    policies, 65
    single phase, 66
    time scales, 65
    times, 3
    workplace, 4, 15, 63, 70

Fairness, 108, 129, 156
    EV charging, 70
    routing, 43
Feedback, 143, 145
Filter, 2, 8, 143
First come first served, 129
Flapping instability, 43
Fleet, *see* Vehicle fleet
    cost, *see* Cost, fleet
    management, *see* Management,
        fleet
Forecast, 64, 161
    weather, 163
    wind power, 164
Framework
    stochastic, 122
Fuel, 88

Full electric vehicle, *see* EV, full
Function
    convex, 184
    cost, *see* Cost function
    piecewise linear, 86
    positive real, 210
    strictly convex, 130
    utility, *see* Utility function

G2V, 82
Game theory, 109
Generation, *see* Power plant
Graph, 185
    directed, 19
    dual, 20
    edge, 18
    node, 18
    primal, 20, 37
    probability, 18
    relation to Markov chain, 18
    search algorithm, 35
    strongly connected, 18
    vertex, 18
Green driving, *see* Ecodriving
Grid, *see* Power grid
Grid to vehicle, *see* G2V
GrooveNet, 197

Hardware-in-the-loop, *see* HIL
HIL, 163, 196, 198
Hoeffding's Inequality, 132
Hybrid electric vehicle, *see* EV,
    hybrid

ICE, 1
    human health, 144
Inconvenience, 86
IntelliDrive, 196
Intelligent speed adaptation, 179
Intelligent traffic management, *see*
    Management, traffic
Internal combustion engine, *see* ICE
Internet of things, 94

Junction, 195

Karush-Kuhn-Tucker, *see* KKT
Kemeny constant, *see* Markov chain,
    Kemeny constant
KKT, 132, 208

Large scale, 8, 193
Liabilities, 94
Linear iteration, 206
Linear program, 173
Load
    balancing, *see* Balancing, load
    electrical, *see* Electrical load
    management, *see* Management,
        load
Logistics, 34
Losses, 24, 87
    transmission, 57
Lyapunov, 211
Lyapunov's inertia theorem, 212

Maintenance, 88
Management
    central unit, 66
    demand, *see* Demand side
        management
    energy, 5, 13, 16, 98, 160
    fleet, 116
    load, 15
    power, 83
    strategies, 66
    supply, 58
    traffic, 4, 13, 14
    vehicle sub-systems, 98
Market growth, 113
Markov chain, 17, 49
    relation to graph, 18
    aperiodic, 50
    definition, 17
    homogeneous, 17
    irreducible, *see* Matrix,
        irreducible
    Kemeny constant, 18, 26
    mean first passage time, 18
    MFPT, *see* Markov chain, mean
        first passage time

Matrix
  column stochastic, 207
  irreducible, 18, 20, 50
  Perron eigenvector, *see* Perron
    eigenvector
  primitive, 18, 20
  transfer function, 211
  transition, *see* Transition matrix
Maximization
  utility, *see* Utility, maximization
Mean first passage energy, 25, *see*
    also Markov chain, mean
    first passage time
Mean first passage time, *see* Markov
    chain, mean first passage
    time
Minimization problem, 130
Mobility, 1, 7, 21
  patterns, 100, 103
  simulator, *see* Traffic simulation
Mode notification, 84
Model
  average-speed, 171
  binomial distribution, 105
  char sharing and queueing, 114
  constant speed, 22
  derivative, 21
  financial, 8, 109
  fixed number of days, 97
  mathematical, 104
  multi-variate, 21
  on-demand, *see* On-demand,
    model
  pollution, 145
  power consumption, *see* Power
    consumption, model
  pricing, *see* Pricing, model
  queueing, 97, 106
  range anxiety, 111
  range of vehicle sizes, 112
  shared-ownership, 93
  sharing, 116
  statistical, 108
  traffic, *see* Traffic model
  user behavior, 108

vehicle access, 97
weighted average, 113
Multiplicative decrease, 206

Network dimension, 209
Network resource allocation, *see*
    Resource, allocation,
    network
Networks
  analogy to mobile cellular
    network, 46
  charge point, 45
  EV, 5, 9
  road, 13, 26, 33, 195, 198
  urban, 21
Normal distribution, *see*
    Distribution, normal
NTS, 101
Nyquist, 212

On-demand
  access, 97
  model, 8, 97
On-off charging, *see* EV charging,
    on-off
Optimization, 88, 162, 169, 173, 183,
    205, 207
  consensus, *see* Consensus,
    optimization

Parking lot, *see* Parking spot
Parking space, *see* Parking spot
Parking spot, 4, 119
  private, 120
  utilization, 125
Passivity, 210
Peak times, 57
Perron eigenvector, 26
Perron-Frobenius theorem, 18, 207
Piecewise linear function, *see*
    Function, piecewise linear
Plug-and-play, 46, 48
Plug-in electric vehicle, *see* EV,
    plug-in

Plug-in hybrid electric vehicle, *see*
    EV, plug-in hybrid
Plug-in-bus, *see* EV, plug-in bus
Policy, 94, 114
Pollution, 7, 86
    aggregate levels, 87, 143, 145
    cooperative control, 147
    fleet, 116
    ICE by-products, 144
    limits, 144
    model, *see* Model, pollution
    regulation, 143
    urban, 1, 145
Positive real, 210
Power consumption, 63
    aerodynamics, *see*
        Aerodynamics, EV power
        consumption
    ancillary systems, *see* Ancillary
        systems, EV power
        consumption
    drivetrain, *see* Drivetrain, EV
        power consumption
    EV, 180
    model, 180
Power demand, 57
    daily demand, 64
Power generation, *see* Power plant
Power grid, 15, 57, 141, 164
    utilization, 57
Power limit, 83
Power management, *see*
    Management, power
Power plant, 88
Power quality, 4, 57
Prediction, 108
Pricing
    model, 99, 113
    real time, 63
    signal, 107
Privacy, 48, 60, 66, 94, 108, 120, 121,
    131, 193, 208
    AIMD, 174, 210
    data, 62
Probability, 67, 106, 123, 195

    tail, 127
Python, 199

Quality of service, 49, 50, 65, 99, 100,
    104, 106, 116, 117, 120, 122
Queueing
    model, *see* Model, queueing
Queueing theory, 49
Queuing, 14
    at charge point, 14

Random variable, 105, 122
    Bernoulli, 124
Random walk, 39
Range, *see also* EV, range, 179
    driving, 116
Range anxiety, 4, 14, 97, 98, 110–113
Reactive power, 63, 82
Real time, 193
Regenerative braking, 24
Regulation, 9
Reinforcement learning, 109
Renewable energy, 81, 160
Renewable source, 57
Resource, 119, 120
    allocation, 122, 145, 205, 207
        network, 145
Restricted areas, 87
Robustness, 48
Routing, 13, 14, 26, 143
    balance load, 45
    charging events, 34
    closed-loop, 43
    collaborative, 15, 41
    collaborative under feedback, 41
    congestion avoidance, 36
    driver mistakes, 37
    energy-aware, 98
    EV, 33
    flapping, *see* Flapping instability
    minimum energy, 33
    minimum fuel, 33
    minimum time, 33
    most economic, 33
    multi-agent, 34

risk averse, 37
selfish, 35
shortest path, 33

Scalability, 48
Scale, 193
large, *see* Large scale
Scale free, 210
Security, 48
Sharing, 9, 103
car, 99, 104
charge point, 119, 120, 132
economy, 5, 93
model, *see* Model, sharing
negotiation, 93
on campus, 120
opportunistic, 93
parking, 119
product, 94
Simulation, 195
traffic, *see* Traffic simulation
Size, 3
vehicle, 97, 110
Smart cities
EV, 141
Smart city, 120
Smart traffic management, *see*
Management, traffic
Speed advisory system, 179
Speed limit, 13, 27, 143
SPONGE, 159, 163
Stability, 211
Standard deviation, 105
Storage device, 16
EV, 58
SUMO, 27, 29, 153, 195
SumoEmbed, 199
Supply management, *see*
Management, supply
Supply-demand balance, 93
Switching system, 185

Team, 159
Time of use, 63
Tires

EV power consumption, 180
Traffic lights, 13
sequencing, 28, 143
Traffic Load, 26
Traffic management, *see*
Management, traffic
Traffic model, 13, 14, 21
energy consumption model, 21
EV, 17
Markovian, 19
Traffic simulation, 195, 196
Transfer function, 210
matrix, *see* Matrix, transfer
function
Transition matrix, 20, 21, 27, 39
twinLIN, 167

Umweltzone, 144, 179
Unintended consequences, *see* V2G,
unintended consequences,
144
Utility
maximization, 9, 151
routing, 43
Utility function, 87, 163, 170, 172,
207

V2G, 16, 58, 63, 81
power flows, 83
unintended consequences, 81, 86
V2I, 169
V2V, 34, 144
connectivity, 7
V2X, 144
Vehicle
gateway, 199
networked, 148
types, 7
virtual, 198
Vehicle access, 104
Vehicle fleet
EV, 179
shared, 109
Vehicle size, *see* Size, vehicle
Vehicle technologies, 159

Vehicle to grid, *see* V2G
Vehicle to vehicle, *see* V2V

Waiting time, 50
Waste materials, 88
Weather forecast, *see* Forecast,
     weather